STUDIES IN PHILOSOPHY

Edited by
ROBERT NOZICK

A ROUTLEDGE SERIES

STUDIES IN PHILOSOPHY
ROBERT NOZICK, *General Editor*

REFERENTIAL OPACITY AND MODAL LOGIC

Dagfinn Føllesdal

Routledge
Taylor & Francis Group
New York London

Published in 2004 by
Routledge
605 Third Avenue,
New York, NY 10017

Published in Great Britain by
Routledge
2 Park Square, Milton Park,
Abingdon, Oxon, OX14 4RN

Routledge is an imprint of the Taylor & Francis Group, an informa business

Library of Congress Cataloging-in-Publication Data
Føllesdal, Dagfinn.
    Referential opacity and modal logic / by Dagfinn Føllesdal.
        p.   cm. — (Studies in philosophy)
    Includes bibliographical references.
    ISBN 0-415-93851-1
    1. Modality (Logic) I. Title. II. Studies in philosophy (New York, N.Y.)

BC199.M6 F6 2002
160 — dc21   2002017877

Publisher's Note
The publisher has gone to great lengths to ensure the quality of this reprint but points out that some imperfections in the original may be apparent.

ISBN 13: 978-0-415-99844-4 (pbk)
ISBN 13: 978-0-415-93851-8 (hbk)

Contents

Introduction

The main reason for publishing this dissertation now, more than forty years after it was written, is historical: it contains what seems to be the first presentation and discussion of the aggregate of ideas that was revitalized a decade later by Geach, Donnellan and Kripke and has come to be known as the New Theory of Reference. For this reason, the text of the original dissertation has been left entirely unchanged. However, some improvements from 1963 have been appended as an addendum.

The basic ideas of the dissertation are as follows:

(1) **Naming**: The dominant Frege-type semantics is given up in favor of a semantics where names and other properly referring expressions keep their reference in "all possible worlds."

Names thereby behave quite differently from other expressions—such as definite descriptions in most of their uses, general terms and sentences—which may change their extension from world to world.

(2) **Identity**: Identity is argued to be universally *substitutive*, and identity statements *necessary*.

(3) **Essentialism**: "Aristotelian" essentialism is defended and shown to be required not only for the modalities, but also for such notions as knowledge, belief, obligation, causality and probability.

(4) **"Slingshot"**: An argument used by Church, Gödel, Quine, and others is seen to lead to unacceptable conclusions and it is shown that the new view on reference blocks this argument.

These are not independent points. A main purpose of the dissertation is to show that they are closely connected with one another and are interrelated parts of an integrated view on reference.

Background: Quine

This essay was submitted as a Ph.D. thesis to the Department of Philosophy, Harvard University, on April 3, 1961. My thesis advisor was W. V. Quine.

Quine was the occasion for my coming to Harvard. He attracted me to Harvard as one of the few philosophers who could think and write clearly about issues I regarded as deep and important.

My plan was to write on reference. When Quine's *Word and Object* came out in 1960 there was, in particular, one issue that struck me as problematic: Quine presented an argument to the effect that when one quantifies into modal contexts, modal distinctions collapse. This argument represented the apparently conclusive culmination of a series of objections Quine had raised against the modalities from 1941 on. His earliest criticism merely demonstrated the obscurity of the modal notions. This new criticism was a step-by-step argument to the effect that the interpretation of the quantifiers and other singular terms forced the modal distinctions to collapse.

I was disturbed by this argument. Not because I cherished the modalities—I did not—but because an inspection of the argument showed that it was independent of the modalities and applied to any attempt to single out by help of an operator a proper subclass of the true sentences. Any such attempt would fail; the subclass would coincide with the whole class. Take as an example "knows that": "There is something that I know to be a book", or "There is somebody whom I know to be a spy" would within a full semantics with names and other definite singular terms lead to a collapse of epistemic distinctions: not only would everything that is known be true, which is what we want, but also everything that is true would be known to be true. Quine's argument could be repeated for belief, causality, counterfactuals, probability, and the operators in ethics, such as "it is obligatory that", "it is permitted that". They would all collapse. The argument was simply too disastrous to be correct.[1]

The problem was then: what is wrong with the argument? In order to find out, I formalized the argument, so that it was turned into a pure deductive argument, the long deduction on page 70 of the dissertation.[2] Often formalization makes one see immediately what is wrong with an argument, by bringing out tacit assumptions that are clearly false. Not so in this case. One should not expect that Quine should have overlooked any dubious assumptions, and all the assumptions he used in his argument were generally accepted at that time. However, something had to be wrong.

The only assumption I could think of questioning, was a transition that is made in the argument between *singular terms* and *general terms*. This is based on a presumption that had gone unquestioned until then, namely that singular and general terms have the same kind of semantics: they have a sense—a descriptive content—and their reference or extension is determined by this sense. This had been taken for granted by all the main figures in the development of semantics, the few indications to the contrary, such as Mill's "Dartmouth," had been forgotten. Frege treated all expressions, singular terms, general terms and sentences, on a par, and he took this unification as an argument in favor of his semantics. Carnap and all twentieth-century semanticists had followed Frege in this, with the exception of Russell, Arthur Smullyan, and a few others who regarded names as a spurious kind of expression that should be eliminated by help of definite descriptions. Russell exempted 'this' from this and called it a "genuine" referring expression (p. 76 below), a label I took over for terms that preserve their reference. By not

---

[1] One could try to prevent the collapse by forbidding quantification into the scope of such operators from outside. However, such restrictions would tend to be *ad hoc*: they would not give any independent reason for forbidding such quantification.

[2] The formalization, and the fact that I discuss the problems of referential opacity by help of formal systems of modal logic, is not primarily due to an interest in formal systems; the issues I discuss are issues we face in our ordinary use of language, and the solutions I propose are supposed to apply to ordinary language. The formalization is only meant as a help to see the problems more clearly, and to express my solutions to them more precisely and concisely. In the case of Quine's argument, formalizing it forced all assumptions of the argument out into the open.

recognizing any names—expressions that refer and are subject to rules of inference such as substitutivity of identity, existential generalization and universal instantiation—Russell and Smullyan evaded the issues of interpretation that concern us in this dissertation. However, these issues remain challenges if one wants to understand how reference works. I believe this is what underlies Gödel's remark in his article on Russell in 1944, to which I shall return below: "I cannot help feeling that the problem raised by Frege's puzzling conclusion has only been evaded by Russell's theory of descriptions and that there is something behind it which is not yet completely understood."[3]

When Quine set forth his criticism, nobody had proposed a system of quantified modal logic that contained names, and the belief that all problems connected with reference were adequately dealt with by Russell's theory of descriptions does a lot to account for the fact that modal logicians were not as much troubled by Quine's arguments as I was.

Names keep their reference in "all possible worlds": Genuine singular terms

Inspection of Quine's argument showed that there was hardly any other way out: one had to give up Frege's and Carnap's neat one-sorted semantics in favor of a two-sorted one, where singular terms behave radically different from general terms and sentences.

The solution could hence only be that names preserve their reference. Or, as I put it in section 17, page 75:

> This solution leads us to regard a word as a proper name of an object only if it refers to this one and the same object in all possible worlds.[4]

---

[3] Kurt Gödel, "Russell's Mathematical Logic." In P.A. Schilpp, ed., *The Philosophy of Bertrand Russell*. Evanston, Ill.: Northwestern University, 1944, pp. 125-53, esp. pp. 128-131 (= pp. 122-124 of the reprint in Solomon Feferman et al., eds., *Kurt Gödel, Collected Works*. Volume II, Oxford: Oxford University Press, 1990, pp. 119-141).

[4] Two years later, in 1963, I proposed and discussed a weaker condition, which only requires reference to the same object in all possible worlds where the object exists. See Addendum, p. 120 ff.

Not all expressions that traditionally have been regarded as singular terms relate in this way to their object. For example, definite descriptions in most of their uses do not. Those expressions that do, I called *"genuine singular terms,"* they are defined in exactly the same way as Kripke in 1970 defined *'rigid designators'*, a very apt phrase for this kind of unswerving expressions.[5] Definite descriptions and other expressions that have traditionally been classified as singular terms, but show philandering behavior, I classified semantically together with general terms. They should be regarded as general terms that just happen to be true of a unique object; they should not be regarded as *referring* to it.

The variables of quantification are the archetypical kinds of genuine singular terms. The whole point of a variable, as of a pronoun, is to refer to an object and keep on referring to that same object, in all possible worlds, through all changes, etc. This is how they are interpreted in this dissertation (section 11 and later sections). One main aim of the dissertation is to show that names behave the same way (although they differ from pronouns and variables in other ways, for example in the way they get attached to an object and how long they keep that attachment).

This conclusion accords well with the analysis of the semantic behavior of referring expressions in section 12. Hence this view on proper names is supported by two kinds of arguments: It blocks Quine's collapse argument, and it results from a semantic analysis of how reference works. The view on reference and names that is proposed in this dissertation is hence not an *ad hoc* obstruction for Quine's argument, but it provides a coherent semantics which makes clear how reference works in contexts that create difficulties for the traditional Fregean view on meaning. This is reflected in the fact that Quine changed his view on the semantics of the modalities after he had read the dissertation, and in his statement in the Foreword to the second, revised edition

---

[5] Kripke presented these ideas in three lectures on reference in Princeton in January 1970, and also in an article, "Identity and Necessity," in Milton Munitz, ed., *Identity and Individuation* (New York: New York University Press, 1971, pp. 135-164). The Princeton lectures were published under the title "Naming and Necessity" in Donald Davidson and Gilbert Harman, eds., *The Semantics of Natural Language* (Dordrecht: Reidel, 1972, pp. 253-355) and later expanded into a book: *Naming and Necessity* (Cambridge, MA: Harvard University Press, 1980).

of *From a Logical Point of View* writes that thanks to this dissertation "the situation has further clarified itself."[6]

The "Slingshot": Gödel's suspicions

Quine's argument for the collapse of modal distinctions is similar to arguments that have been used by Church and Gödel for the collapse of various other distinctions. Church[7] used an argument of this kind to argue for his Fregean semantics. Gödel[8] proposed an even simpler argument and showed that it "leads almost inevitably to the conclusion that all true sentences have the same signification (as well as all false ones)." (pp. 129-30) He noted that "Frege actually drew this conclusion; and he meant it in an almost metaphysical sense, reminding one somewhat of the Eleatic doctrines of the 'One'." (p. 129). Gödel presented the argument in a discussion of Russell's philosophy, and he observed that Russell's contextual elimination of definite descriptions blocks the argument. However, Gödel noticed that the argument raised serious problems for the traditional Fregean view on names and reference. He concluded, as noted above, that "I cannot help feeling that the problem raised by Frege's puzzling conclusion has only been evaded by Russell's theory of descriptions and that there is something behind it which is not yet completely understood" (p. 130).

A hint of what Gödel had in mind here, comes a little bit later, where he writes that: "Closer examination, however, shows that this advantage of Russell's theory over Frege's subsists only as long as one interprets definitions as mere typographic abbreviations, not as introducing names for objects described by the definitions, a feature which is common to Frege and Russell" (p. 131).

So what Gödel was missing, was a semantics for names that gets around his argument. In this dissertation, I propose such a way of handling names: I argue

---

[6] W. V. Quine, *From a Logical Point of View*, Second edition, revised, New York: Harper Torchbooks, 1961. Foreword, p. vi.

[7] Alonzo Church, Review of Carnap's *Introduction to Semantics. Philosophical Review* 52 (1943), pp. 298-304, esp. pp. 299-300.

[8] Kurt Gödel, *op.cit*, pp. 125-53, esp. pp. 128-131 (= pp. 122-124 of the reprint in Solomon Feferman et al., eds., *Kurt Gödel, Collected Works*. Volume II, Oxford: Oxford University Press, 1990, pp. 119-141). The page numbers in the text refer to the original text.

that names do not philander from object to object the way definite descriptions do, but that they stick to the same object in all possible worlds. I mention as an alternative that eliminating all definite singular terms by way of descriptions would get us around the argument. This alternative has the advantage of simplicity. However, names in ordinary language do not behave like disguised descriptions. On page 75, where I write "this solution leads us to regard a word as a proper name of an object only if it refers to this one and same object in all possible worlds," I continue:

> This does not seem unnatural. Neither does it seem preposterous to assume as we just did, that if a name-like word does not stick to one and the same object in all possible worlds, the word contains some descriptive element.

As far as I know, Gödel was the only one who had regarded the argument with suspicion before it was discussed critically in this dissertation. My diagnosis is, as mentioned, that in the argument one slides too easily between general terms and singular terms. The proposed two-sorted semantics prevents such a slide and does justice to the behavior of names in natural languages. In the years that have followed, a large number of articles and books have been devoted to the argument, often repeating earlier work. A notable contribution was made by Barwise and Perry in 1981.[9] The very apt name "the slingshot" is due to them.

---

[9] Jon Barwise and John Perry, "Semantic Innocence and Uncompromising Situations." *Midwest Studies in Philosophy* 6 (1981), pp. 387-404. See also John Perry, "Evading the Slingshot." In A. Clark, et al., eds., *Philosophy and Cognitive Science*. Dordrecht: Kluwer, 1996, pp. 95-114. Reprinted in John Perry, *The Problem of the Essential Indexical and Other Essays*. Expanded Edition, Stanford: CSLI Publications, 2000, pp. 287-301. For further discussion, see my "Situation semantics and the 'slingshot' argument." In C. G. Hempel, H. Putnam, and W. K. Essler, eds., *Methodology, Epistemology, and Philosophy of Science. Essays in Honour of Wolfgang Stegmüller on the Occasion of His 60th Birthday*, June 3, 1983. *Erkenntnis* 19 (1983), pp. 91-98. See also Stephen Neale, *Facing Facts*. Oxford: Oxford University Press, 2001, and the literature referred to there.

Substitutivity of identity and necessity of identity

During the twenty years before the publication of *Word and Object* Quine had directed a number of other arguments against the modalities. Many of these survive after one has introduced a two-sorted semantics. When analyzed against the background of this new semantics, they turn out to be innocent. Rather than being objections they give insight into how the two-sorted semantics works. This holds for Quine's claim that identities are universally substitutive and necessary, and also for his contention that quantified modal logic commits one to "Aristotelian" essentialism.

Let us consider identity first. In section 12, I argue that identities are necessary (Thesis I). Non-identity, or distinctness, is, however, not necessary (section 14). In section 17, I argue for universal substitutivity of identity and in section 19, I show how the various difficulties connected with singular terms are resolved within two-sorted semantics (pp. 87 ff.).

In particular, the substitutivity of identity has disturbed many. Already in his first criticism of the modalities, his contribution to the Whitehead volume in the *Library of Living Philosophers* in 1941,[10] Quine observed that the principle of substitutivity of identity seems to break down in modal contexts: If in '$\Diamond$ (the number of planets < 7)' which is presumably true, we substitute for '9' 'the number of planets', we get '$\Diamond$ (9 < 7)', which is false. So what, then, is meant by the identity statement 'the number of planets = 9'?

This objection was utterly distressing within traditional one-sorted semantics, à la Frege and Carnap. Some modal logicians tried to get around it by contending that names and variables in modal contexts refer to intensions rather than to regular objects. However, it is argued in this thesis that restricting the universe to intensions is the wrong way to go. The trouble lies with the referring expressions and their semantics, not with the objects:

> The variables and quantifiers by themselves do not require us to individuate our entities finer than they are individuated in extensional logic. The trouble comes in with our singular terms. (Section 17, page 74)

---

[10] "Whitehead and the rise of modern logic." In Paul Arthur Schilpp, ed., *The Philosophy of Alfred North Whitehead* (Library of Living Philosophers). Evanston, IL: Northwestern University, 1941. 2.ed. New York: Tudor, 1951, pp. 125-163, esp. pp. 141-142, n. 26, and p. 148.

Others, for example Ruth Marcus, have suggested substitutional interpretations of quantifiers and names.[11] However, this is a radical deviation from normal quantification and reference, and it throws no light on the relation between language and the world. Others again have introduced restrictions on sub-stitutivity and have maintained that not all singular terms can be substituted for one another in modal contexts. However, no argument has ever been given for such a restriction, except that it saves one from undesirable results, such as those pointed out by Quine.

Before 1961 nobody had ever suggested giving up the old Frege-Carnap view on reference in favor of a new view on reference, where names and other "genuine singular terms" refer, while other expressions that have traditionally been regarded as singular terms, including definite descriptions in most of their uses, are classified semantically together with general terms and sentences, as having a sense that determines their extension. Given this new semantics, statements like Quine's 'the number of planets = 9' are not regarded as proper identity statements, the description on the left side is not a referring expression, but is a general term that in the actual world happens to be true of the number 9. (See section 19, pages 87 ff.)

Two-sorted semantics

Such a two-sorted semantics, where general terms and sentences behave as in standard Fregean semantics, while singular terms keep their reference "in all possible worlds," undermines Quine's argument. There is no collapse of modal distinctions, and quantification into modal contexts makes sense. This is due to the fact that two-sorted semantics makes it possible to have contexts that are referentially transparent, so that quantification into them makes sense, and extensionally opaque, so that modal distinctions do not collapse. As I argue in section 5, not all combinations of referential transparency and extensional opacity are permissible. Thus, for example, one may prove that:

> *Every extensional construction on sentences and general terms is referential*

---

[11] For example in "Modal Logics I: Modalities and Intensional Languages." *Synthese* 13 (1961), pp. 303-322. This paper will be discussed below.

and likewise that:

Every referential construction on singular terms is extensional

However, luckily for quantification into non-extensional contexts:

Referential constructions on sentences and general terms can be non-extensional

The constructions we are discussing, where the operators 'necessarily', 'possibly', 'knows that', 'believes that', 'it is probable that', 'it is obligatory that', etc, apply to sentences or general terms, are just such constructions.

"Aristotelian" essentialism

Another of Quine's points against modal logic which many have objected to is his contention that quantified modal logic presupposes "Aristotelian" essentialism. This is the doctrine that some of the attributes of a thing are essential to it, necessary of the thing *regardless of the way in which we refer to it,* while other attributes are accidental to it. This doctrine may have little to do with Aristotle's position, and the label "Aristotelian" may therefore be misleading. What matters is that Quine has defined a notion of essentialism that is highly relevant to semantics. It is argued in this dissertation that "Aristotelian" essentialism is indispensable in quantified modal logic. However, while Quine's objection was originally intended as a *reductio* of modal logic, it is concluded in this dissertation that

> if the modal operator '□' itself makes sense, then . . . open sentences with a '□' prefixed make sense too if we restrict our stock of singular terms appropriately.
> To make sense of Aristotelian essentialism and to make sense of open sentences with a '□' prefixed are one and the same problem, and a solution to the one is a solution to the other. (Section 19, page 92)

The combination of extensional opacity and referential transparency which is made possible by two-sorted semantics is just what "Aristotelian" essentialism amounts to:

> We distinguish between necessary and contingent attributes (*extensional opacity*), and the objects over which we quantify have these attributes regardless of the way in which the object is referred to (*referential transparency*).

Carnap and other advocates of modal logic who explained modality by help of analyticity and who rejected Aristotelian essentialism as metaphysical nonsense might find this result hard to swallow. However, Quine himself changed his attitude to Aristotelian essentialism. On February 8, 1962, in a discussion following a lecture by Ruth Marcus, he said: "I think essentialism, from the point of view of the modal logician, is something that ought to be welcome. I don't take this as being a *reductio ad absurdum*."[12]

There are many misunderstandings as to what Aristotelian essentialism amounts to. First, it does not mean that every object has an individual essence which is unique to it and distinguishes it from all other objects. Nor does it mean that for each kind of object there is some essence that all things of that kind have. Further, and this is important for understanding the semantic issues, it does not mean that each object, or perhaps each named object, has some essential trait by which we can pick it out. As can be seen from the arguments that show that essentialism is needed, at the end of section 10 and in the arguments leading up to section 19 of the dissertation

> *"Aristotelian" essentialism has nothing to do with names, it is required already in order to interpret the quantifiers and variables, where no names are involved.*

Unfortunately, Quine himself has contributed to this misunderstanding. In "Intensions Revisited" (1977), the only place where Quine addressed these issues at any length after the revised edition of *From a Logical Point of View* in the fall of 1961, Quine noted that in order to make sense of quantified modal logic one needs a kind of singular term 'a' for which one can establish '(Ex)□(x = a)'. He then goes on to say:

---

[12] "Discussion after Ruth Barcan Marcus's lecture February 8, 1962." In Max Wartofsky, ed., *Boston Studies in the Philosophy of Science*. Dordrecht: Reidel, 1963, pp. 105-116. The quoted passage occurs on page 110.

> A term thus qualified is what Føllesdal called a genuine name and Kripke has called a rigid designator. It is a term such that $(Ex)\Box(x = a)$, that is, something is necessarily a, where 'a' stands for the term.
>
> . . .
>
> A rigid designator differs from others in that *it picks out its object by its essential traits*. It designates the object in all possible worlds in which it exists. Talk of possible worlds is a graphic way of waging the essentialist philosophy, but it is only that; it is not an explication. *Essence is needed to identify an object from one possible world to another.*[13]

The first paragraph characterizes genuine singular terms, or rigid designators, the way they should be characterized, and have been characterized above. However, the second paragraph introduces a notion of essence that is not supported by arguments and is not called for in modal logic. It might be helpful to distinguish two different notions of essentialism. The first, weak notion was developed by Quine in response to Carnap, Lewis, and others, who championed quantified modal logic while at the same time rejecting as metaphysical nonsense the traditional Aristotelian view that necessity inheres in things and not in language. In *From a Logical Point of View* Quine says that quantification into modal contexts requires Aristotelian essentialism, in the following sense:

> An object, *of itself and by whatever name or none*, must be seen as having some of its traits necessarily and others contingently, despite the fact that *the latter traits follow just as analytically from some ways of specifying the object as the former traits do from other ways of specifying it*.
>
> . . .
>
> Essentialism is abruptly at variance with the idea, favored by Carnap, Lewis, and others, of explaining necessity by analyticity.[14]

---

[13] *Midwest Studies in Philosophy* 2 (1977), pp. 5-11. Reprinted in Quine's *Theories and Things*. Cambridge, MA: Harvard University Press, 1981, pp. 113-123, where the quoted passage appears on page 118, the italics are mine.

[14] *From a Logical Point of View*. Cambridge, MA: Harvard University Press, 2. ed. 1961, p. 155. The italics are mine. The passage does not occur in the first edition.

Quine saw that Carnap and Lewis's linguistic conception of necessity was untenable if one wants to quantify into modal contexts, and that their position therefore was incoherent.

The second, strong notion of essentialism is evoked by a very different historical situation: the discussion amongst modal logicians in the seventies concerning how one should identify objects from one possible world to another, how one should draw what David Kaplan appositely dubbed "trans-world heir lines". One proposal that sometimes was made, was that each object had an individual essence, a set of properties that it had by necessity and that no other object had. It is clear from the last italicized clause in the quotation above from Quine's 1977 paper that it is this issue Quine addresses here, and no longer those raised by Carnap and Lewis.

It seems to me that Quine in 1961 was satisfied that he had refuted Carnap and Lewis's views on the modalities and had shown that quantified modal logic requires what he then called "essentialism". He also acknowledged that the formal difficulties that he had brought to light in his writings on quantified modal logic could be overcome if one introduced a notion of "genuine" singular terms, and in the summer and fall of 1961 he rewrote, as mentioned above, the sections on modality in *From a Logical Point of View* to reflect the clarified situation.

When some modal logicians in the seventies argued that there was a problem of identifying an object from one possible world to another and appealed to a notion of essence for this purpose, Quine seems to have regarded this as just another manifestation of his old claim that quantified modal logic requires essentialism. However, in my view, Quine was wrong here. The problem of trans-world identification is an ill-conceived problem; it is based on a misconception of how genuine singular terms work, and the strong notion of essence that has been introduced in order to take care of such identification is not required for modal logic, while the weaker kind, which Quine discussed in his earlier writings, is needed.

Although I think that Quine was fully right when in 1977 he expressed his misgivings about trans-world identity and about the strong notion of essence invoked to account for it, I find the quoted passage unfortunate in two ways: First, it may make the reader confuse the subtle arguments concerning quantified modal logic and essentialism that Quine gave in the forties and fifties and that refuted Carnap and Lewis, with the trifling claim that essences are needed to identify objects from one possible world to another. Secondly, the passage gives a wrong view on the nature of reference, as epitomized by the behavior of genuine singular terms.[15]

Much of the later discussion of essentialism, by Ruth Marcus, Terrence Parsons and others, is marred by confusion on this issue. They argue against Quine that modal logic does not require essentialism. However, the essentialism they discuss is not the essentialism that Quine's arguments show to be required by modal logic.

Carnap

The dissertation contains, in section 13, pp. 46-48, a discussion of the semantics of Carnap's systems MFC in Modality and Quantification (19) and his system S_2 in *Meaning and Necessity*. The first systems of quantified modal logic that were proposed had no singular terms other than variables. Since variables keep their reference from one possible world to another, the collapse discussed by Quine was not brought to the fore until one got systems of quantified logic that included singular terms other than variables. The only such system by 1961, when this dissertation was written, was Carnap's S_2. It contained definite descriptions, but no names. In the discussion of definite descriptions in section 16 of the dissertation it is proved, on pages 73-73, that:

> S_2 is saved from a collapse of modal distinctions by the circumstance that its theory of descriptions is not a standard one [Carnap states on page 184 of *Meaning and Necessity* that "in order to avoid certain complications, which cannot be explained here, it seems advisable to admit in S_2 only descriptions which do not contain '□'].[16]

---

[15] For a fuller discussion of this, see my "Essentialism and reference." In Lewis E. Hahn and Paul Arthur Schilpp, eds., *The Philosophy of W. V. Quine* (The Library of Living Philosophers). La Salle, IL: Open Court, 1986, pp. 97-113. (Reply by Quine: pp. 114-115).

[16] A thoughtful attempt to rescue Carnap has been made by Genoveva Marti in "Do Modal Distinctions Collapse in Carnap's System?" *Journal of Philosophical Logic* 23 (1994), pp. 575-594. I think Marti is right in her most interesting analysis of Carnap's motivation for his seemingly so *ad hoc* prohibition of modal operators within definite descriptions. However, her attempt to block my argument for the collapse turns in my opinion too strongly

Church

In an Appendix to the dissertation (pp. 115-118) I discuss a system of intensional logic that Alonzo Church proposed in "A formulation of the logic of sense and denotation" (1951).[17] In 1943, in a review of Quine's "Notes on Existence and Necessity", Church had argued that one can quantify into modal contexts, provided the quantifier has an intensional range—a range, for instance, composed of attributes rather than classes. For ten years, until 1953, Quine accepted this. However, he then came to see that restrictions on one's universe of discourse are of no avail.

In the meantime, Church had worked out his "logic of sense and denotation." What saves this system from collapse is not restrictions on the universe of discourse, but a Frege-inspired reference shift that takes place within modal contexts: what object a variable takes as value, depends on the modal operators (or in Church's case, rather modal predicates) within whose scope it occurs. Thanks to this feature, Church's system does not have any opaque contexts. All its contexts are referentially and extensionally transparent, and there is no need to have a two-sorted semantics in order to prevent a collapse of modal distinctions. Although it uses symbols like 'N', for 'is necessary', Church's system is not a system of modal logic; it is a purely extensional system. It opens an interesting alternative to modal logic that has not yet received the attention it deserves.

---

on the unclarities in Carnap's double interpretation of variables, which I criticize in this dissertation.

[17]Alonzo Church, "A Formulation of the Logic of Sense and Denotation," in Paul Henle, H. M. Kallen and S. K. Langer (eds.) *Structure, Method and Meaning: Essays in Honor of H.M. Sheffer*, Liberal Arts Press, New York, 1951, pp. 3-24. (Abstract in *Journal of Symbolic Logic* 11 (1946), p. 31.) An improved version has later been published as "Outline of a Revised Formulation of the Logic of Sense and Denotation." Part I, *Noûs* 7 (1973), pp. 24-33; Part II, *Noûs* 8 (1974), pp. 135-156.

Some logical theorems

As remarked above, although this dissertation makes use of formalism and discusses various systems of modal logic, its aim is to discuss philosophical issues that come up in natural languages, in particular to get a grip on the notion of reference which is so crucial for our understanding of the relation between language and the world. The formalization is only meant as a help to see the problems more clearly, and to express my solutions to them more precisely and concisely.

As a side product, the dissertation contains some simple contributions to modal logic. The purely formal results in the dissertation are only incidental to the philosophical discussion. In view of the semantic analysis some systems of modal logic are proved to be semantically incomplete; in addition to Carnap's S_2, quantifier-identity extensions of six of the eight Lewis systems (Theorems III and V-XII). Some theorems are proved that have been proved independently by others later, mainly theorems dealing with the rule of inference RL (if $\vdash \varphi$ then $\vdash \Box\varphi$) and with identity and distinctness of individuals and with mixtures of quantifiers and modal operators, such as the Barcan formula '$(x)\Box Fx \supset \Box(x)Fx$' and its converse . Most of these proofs are found in sections 13 and 15.

A routine part of the dissertation that some might find useful, is section 9, which gives a survey of the different systems of quantified modal logic that had been proposed up to 1961, and section 10, which reviews the various objections that had been raised against the modalities.

The dissertation contains, in section 11 (pp. 29-33), the first systematic presentation of the basics of what later has been called "Kripke semantics." No originality is claimed for these ideas; as I point out at the beginning of this section (footnote 26, page 30), I was indebted to conversations with Saul Kripke for the interpretation of iterated modalities that was outlined in that section. Kripke had at that time, in the spring of 1960-61, worked out most of the approach to iterated modalities that he presented in 1963.[18] It is important here to distinguish so-called "Kripke semantics," which has to do with the interpretation of the modalities, especially iterated modalities, from the

---

[18] Saul Kripke, "Semantical Considerations on Modal Logic." *Acta Philosophica Fennica* 16 (1963), pp. 83-94, and "Semantical Analysis of Modal Logic: I, Normal Modal Propositional Calculi." *Zeitschrift für mathematische Logik und Grundlagen der Mathematik* 9 (1973), pp. 67-96.

semantics of names and other singular terms, which is the topic of this dissertation.[19]

Marcus and Kripke

Much attention has been given to the meeting at Harvard in February 1962, which I mentioned above in connection with essentialism. Ruth Marcus there read a paper announced under the title "Foundation of Modal Logic," and it has been claimed that Kripke picked up his idea of rigid designators at this talk.[20] Kripke was at that time fully immersed in his work on modal logic. I suggested to him that he come with me to the meeting, and he reluctantly came along. Kripke has later told me that he regretted this decision very much, since the accusations of plagiarism have affected him strongly. However, when one reads Ruth Marcus's lecture and the transcription of the discussion, there seems to be no basis for the accusation.[21] Marcus suggests that proper names can be regarded as tags, but this idea is not integrated in the rest of the paper. She proposes instead a substitutional reading of the quantifiers. Such a substitutional reading

---

[19] In an article in *Times Literary Supplement* (February 9, 2001, pp. 12-13) Stephen Neale has speculated that these conversations led us to gravitate towards a common position. I would like to mention here that these conversations were exclusively devoted to the interpretation of the modalities, and not to the issues relating to names and other singular terms. Kripke's interest in the latter issues arose only later, in 1963-64 (see Kripke, *Naming and Necessity*, esp. pp. 3 and 5) and led to his three lectures on reference in Princeton in January 1970.

[20] See Paul W. Humphreys and James H. Fetzer, *The New Theory of Reference.* Dordrecht: Kluwer, 1998, esp. pp. viii-ix. This volume contains the 1994 paper where Quentin Smith made this claim and various other papers that defend Kripke or trace the origins of the New Theory of Reference.

[21] Ruth Marcus's paper was later published under the title "Modal Logics I: Modalities and Intensional Languages," *Synthese* 13 (1961), 303-322, followed by Quine's comments (pp. 323-330). The issue is dated December 1961, that is, before the talk, but this is due to an irregularity in *Synthese*'s publication schedule. Marcus's talk, Quine's comments and a transcript of the discussion at the meeting were published in the 1963 volume of *Boston Studies in the Philosophy of Science*, edited by Wartofsky, referred to above.

would not be called for if a theory of proper names had been incorporated in her semantics. Instead, her interpretation of quantified modal logic becomes very different from the one that is proposed in this dissertation (which Ruth Marcus had not read by the time she gave her talk). Also, there are various other unclarities in her conception of names as tags; she says, for example:

> If one wishes, one could say that object-reference (in terms of quantification) is a wider notion than thing-reference, the latter being also bound up with identity and perhaps with other restrictions as well such as spatio-temporal location. If one wishes to use the word 'refer' exclusively for thing-reference, then we would distinguish those names which refer, from those which name other sorts of objects.[22]

This conception of names naming some special kind of objects, different from those that the quantifiers range over, is certainly very far from the conception of names advocated in this dissertation, and also from the views that Kripke has been accused of taking over from Ruth Marcus.

There is also some confusion in Marcus's talk and in the ensuing discussion about the relation of names and essentialism. Marcus, and also Kripke, thought that essentialism has to do with names. Kripke stated this view most clearly:

> It seems to me the only thing Professor Quine would be able to say and therefore what he must say, I hope, is that the assumption of a distinction between tags and empirical descriptions, such that the truth-values of identity statements between tags (but not between descriptions) are ascertainable merely by recourse to a dictionary, amounts to essentialism itself. The tags are the "essential" denoting phrases for individuals, but empirical descriptions are not, and thus we look to statements containing "tags", not descriptions, to ascertain the essential properties of individuals. Thus the distinction between "names" and "descriptions" is equivalent to essentialism.[23]

As noted above, my discussion of essentialism in this dissertation does not have to do with names, essentialism comes in already when we quantify into modal contexts: "To make sense of Aristotelian essentialism and to make sense

---

[22] Page 106 of the transcript in Wartofsky.

[23] Page 115 of the transcript in Wartofsky.

of open sentences with an '□' prefixed are one and the same problem, and a solution to the one is a solution to the other." (Section 19, page 92) This is also what Quine notes in his answer to Kripke in the 1962 discussion:

> My answer is that this kind of consideration is not relevant to the problem of essentialism because one doesn't ever need descriptions or proper names.[24]

It may be that we here have the beginning of the different notions of essentialism that I mentioned above. Also, if names are supposed to refer in virtue of essential properties, then we are back to regarding names as a variety of descriptions again. However, it is not so clear whether this is what Marcus and Kripke have in mind here.

Although I disagree with Marcus's way of treating names and quantifiers here, this does not diminish my indebtedness to her. She is, beside Quine, the person to whom I refer most frequently in this dissertation. She has been one of the main contributors to modal logic and has been a pioneer in many ways: she was the first to propose a system of quantified modal logic, and she has proved and defended theorems like '$(x)(y)(x=y. \supset \Box(x=y))$', which have been contested by many, but which, as I argue in this dissertation, are forced upon us when we reflect on reference, identity and modality.

Later developments

In this dissertation I have tried, as one always should, to mention everybody who has had similar ideas and in particular to identify those who first put forth the various ideas and argued well for them. I would have liked to survey the further development of the subject beyond 1961. However, this would be a major enterprise and go far beyond the scope of this introduction. I can only refer to the bibliographies in the major books on the subject; in particular the books by Neale and by Humphreys and Fetzer that I referred to earlier.

---

[24] Page 115 of the transcript in Wartofsky.

The normative view on reference

One main issue in the later discussion has been how one should conceive of the connection between name and object. What kind of linkage can insure that a name keeps the same reference in all possible worlds? I could conceive of no solution to this problem when I wrote my dissertation. On the one hand names and other genuine singular terms must keep their reference in order for quantification to make sense. On the other hand, history is full of examples of names that due to confusion have come to change their reference. It took me many years to notice something that should have struck me immediately: What I show in this dissertation is not that names and other referring expressions keep their reference in all possible worlds, I show only the conditional statement that *if* quantification into modal (and other intensional) contexts shall make sense, *then* names and other referring expressions have to keep their reference.

We have hence no guarantee that names keep their reference, we only know that *if* we get confused about reference, *then* we get confused about quantification. When we use a name, a pronoun or a quantificational variable, we signal that we intend to keep on referring to the same object, and we commit ourselves to do our best to keep track of it. It may be a deep fact of our relation to the world that we conceive of it as consisting of objects. Objects play important roles in our life; it is often vital to follow them through changes, to learn more about them, to correct our false beliefs about them, etc. Their importance as stabilizing factors in our environment is reflected in language, through the apparatus and the behavior of singular terms.[25]

Constancy of reference is therefore not something which is guaranteed, but something we must strive for when we use singular terms. It is a norm that we

---

[25] See my "Reference and sense." Symposium on reference (together with Kripke, Strawson and Quine) at the XVIIth World Congress of Philosophy, Montreal, August 21–27, 1983. In Venant Cauchy (ed.), *Philosophy and Culture: Proceedings of the XVIIth World Congress of Philosophy*. Montreal: Editions du Beffroi, Editions Montmorency, 1986, pp. 229-239. This is a paper I read at various universities from the beginning of the 1970's; see my discussion of Evans below.

See also my "Essentialism and reference." In Lewis E. Hahn and Paul Arthur Schilpp, eds., *The Philosophy of W. V. Quine* (The Library of Living Philosophers). La Salle, IL: Open Court, 1986, pp. 97-113, and Quine's reply, pp. 114-115.

are expected to live up to as language users. I have therefore called this a *normative* view on reference.

One might wonder what then happens to the whole idea that names are genuine singular terms, or rigid designators, that refer to the same object in all worlds (or in all worlds in which the object exists). Since I regard names as special not because they always succeed in preserving their reference, but because they commit us to do our best to keep track of the reference, I regard this latter, normative, feature of genuine singular terms as their defining characteristic.

Historical chain views on reference

Most of the attempts to analyze the link between names and object have taken it for granted that names succeed in sticking to their reference. Some invoke a *historical chain* through which the name is transmitted, other, so-called causal theories of reference, allege that there is a *causal* relation that provides the tie.

The historical chain theories were the first to be proposed, by Peter Geach (1969),[26] Keith Donnellan (1970),[27] and Saul Kripke (lectures given in January 1970, published in 1972).[28] The following passage from Geach encapsulates his view on the historical chain:

> I do indeed think that for the use of a word as a proper name there must in the first instance be someone acquainted with the object named. But language is an institution, a tradition; and the use of a given name for a given object, like other features of language, can be handed down from one generation to another; the acquaintance required for the use of a proper name may be mediate, not immediate. Plato knew Socrates, and

---

[26] Peter Geach, "The Perils of Pauline." *Review of Metaphysics* 23 (1969), pp. 287-300. Reprinted in Geach, *Logic Matters*, Oxford: Blackwell, 1972, pp. 153-165.

[27] Keith Donnellan, "Proper Names and Identifying Descriptions." *Synthese* 21 (1970), pp. 335-358. Reprinted in Donald Davidson and Gilbert Harman, eds., *Semantics of Natural Language*. Dordrecht: Reidel, 1972, pp. 356-379.

[28] Saul Kripke, "Identity and Necessity," and *Naming and Necessity*, referred to above.

Aristotle knew Plato, and Theophrastus knew Aristotle, and so on in apostolic succession down to our own times; that is why we can legitimately use 'Socrates' as a name the way we do. It is not our knowledge of the chain that validates our use, but the existence of such a chain, just as according to Catholic doctrine a man is a true bishop if there is in fact a chain of consecrations going back to the apostles, not if we know that there is. When a serious doubt arises (as happens for a well-known use of the word 'Arthur') whether the chain does reach right up to the object named, our right to use the name is questionable, just on that account. But a right may obtain even when it is open to question.[29]

There are differences between Geach, Donnellan and Kripke. One is that while Geach requires there to have been direct acquaintance with the object at the beginning of the historical chain through which the name reached us, and Donnellan requires there to be "a historical connection with some individual," Kripke lets the chain begin with an act of baptism, where the object baptized need not be present, but, for example, can be picked out by a definite description. Clearly, Geach's view runs into problems with names for abstract objects and for future objects and events. Donnellan is aware of this problem and refines and develops his view in interesting ways in later articles. It is unfortunate that both Donnellan and Geach have been given far too little attention and credit in later discussions.

Causal theories of reference

The first causal theory was proposed by Gareth Evans in 1973.[30] Evans modestly called his theory a further development and tightening up of Kripke's view, which he dubbed the "Causal Theory." However, Evans's view is importantly different. The gist of his theory is that when a name is used,

---

[29] Pages 288-289 of the original article, page 155 of the reprint.

[30] Gareth Evans, "The Causal Theory of Names." *Aristotelian Society Supplementary Volume* XLVII (1973), pp. 187-208. Reprinted in Stephen P. Schwartz, *Naming, Necessity, and Natural Kinds*. Ithaca: Cornell University, pp. 192-215, and in Gareth Evans, *Collected Papers*. Oxford: The Clarendon Press, 1985, pp. 1-24.

> I think we can say that in general a speaker intends to refer to the item that is the dominant source of his associated body of information.[31]

Evans has a causal theory of information, and his theory, unlike Kripke's, rightly should be called "causal." Shortly after Evans's paper came out I gave a lecture in Oxford called "Reference and Meaning," where I presented the "normative" view on reference that I have sketched above. Since Evans was the commentator on my paper, I used the opportunity to include some criticism of his paper, mainly the following two: First, Evans has the same problems as Geach and Donnellan, in explaining reference to abstract entities or to future entities and events. Presumably, such objects cannot be causes of our information. Secondly, while I think it is valuable to bring in the notion of information, it is far from clear what is meant by 'the dominant source of one's information'. In the discussion, Alfred Ayer illustrated this objection by remarking that according to Evans's theory, the reference of most of the names he learned as a child was his nurse, since she was the dominant source of the information he associated with these names.

One main problem I have with all these theories, Geach's, Donnellan's, Kripke's, Evans's and many that have come later, is that they take the tie between the name and the object to be established by factual matters quite independently of whether we know about them. Geach's apostolic succession theory is very explicit: "It is not our knowledge of the chain that validates our use, but the existence of such a chain, just as according to Catholic doctrine a man is a true bishop if there is in fact a chain of consecrations going back to the apostles, not if we know that there is." Such an infallible historical or causal chain which insures sameness of reference regardless of our mistaken beliefs and confusions is, of course, just what we would want if names have to refer to the same object in all possible worlds. This is part of the appeal of such chains. However, my view is that although the transmission process and changes in our information play an important role in determining what our names refer to, their reference may change. Community-wide confusion and numerous other factors may contribute to such changes. Only a detailed study of actual cases where there have been reference changes, such as in Evans's example of 'Madagascar', can tell us how our commitment to preserve reference works, and how it may be defeated.

---

[31] Page 208 of the reprint in Schwartz.

Given that the use of names and other genuine singular terms commits us to keep on referring to the same object, one needs two elements in a theory of names and reference. First, given the emphasis on the object being the same, one needs a theory of the individuation of objects. Theories of personal identity are an example. Secondly, one needs a theory of how names and other referring expressions relate to the objects they refer to. This is largely an empirical matter, which requires detailed empirical studies of language use in a social setting. As I mentioned earlier, I do not think that essences play any role in either of these issues. Appeal to "individual essences" seems to me to be question-begging with regard to the individuation of objects. And to think that names relate to their objects by help of their essential properties seems even more farfetched. As noted above in connection with the debate after Ruth Marcus's Harvard lecture this also seems to come dangerously close to a revival of the description theory of names.

What is not done in this dissertation

I have already mentioned one main point that is not discussed in this dissertation: the kind of linkage there is between names and their objects. In the dissertation I argue that names are genuine singular terms: they refer to the same object in all possible worlds. This is shown by semantical considerations concerning the interpretation of modalities and quantification. However, I do not discuss various possible views of how names are linked to their object: through historical transmission, through causal links, or in some other way.

Also, I do not argue for my view on the semantics of names by surveying the behavior of names in ordinary language. This is admirably done in Kripke's work. Further, I do not go into the question of how broadly the category of names should be conceived. I argue that any expression that is subject to rules of inference such as substitutivity of identity, existential generalization and universal instantiation should keep its reference in all possible worlds, but I do not discuss whether this comprises names for natural kinds.

Addendum

This dissertation was written in a great hurry in the winter of 1960-61. Two years later, in connection with an application for a job at the University of Oslo,

I rewrote it, improved the exposition, and added a discussion of necessary existence, arguing that the requirement that names and other genuine singular terms have to refer to the same object in all possible worlds was too strong; all that is needed is that they refer to the same object in all worlds where the object exists. I here used Frege's idea of letting formulas be without truth value if they contain names without a reference. The move was inspired by Arthur Prior's modal system Q in *Time and Modality* (Prior 1957). In January 1966 this improved version, with an updated bibliography, was published in mimeographed form by Oslo University Press. The few mimeographed copies were sold out quickly, and the dissertation has therefore largely been accessible only in its original form, on microfilm or by loan from Harvard University Library.

In the addendum I have included the main changes of the mimeographed version, notably the discussion of necessary existence, which affects the end of section 20 and also the many other places where I say that names and other genuine singular terms keep their reference in all possible worlds.

The bibliography has been kept unchanged. The literature on these topics has become immense and a full bibliography would need a separate volume. Selected bibliographies may be found in the book by Humphreys and Fetzer referred to in footnote 20 above.

Acknowledgments

My original dissertation contained no thanks or acknowledgments, since I found it awkward to express what I owed Quine in a dissertation he was going to judge. However, the preface to the mimeographed version of my dissertation contained the following acknowledgements:

I wish to express my gratitude to Professor Quine for his most stimulating teaching and advice while I was writing my thesis. His insight and clarity have been far more important to me than could be expressed by the many references to him in the text.

I also owe a special thank to Mr. Saul Kripke, who in the spring of 1961 imparted to me virtually all the basic ideas on iterated modalities that he presented in his 1963 and 1964 papers. (Cf. section 11, note 26).

My thanks are also due to Professors Burton Dreben, Neil Wilson and Charles Parsons, who have read the manuscript and suggested several improvements, and to my other teachers of philosophy.

Finally, I want to thank the United States Government for a Smidt-Mundt Scholarship, which first brought me to this country, and the University of Oslo, the Norwegian Research Council for Science and the Humanities, and Harvard University for the economic support which they have given me while I studied philosophy and wrote this essay.

I would now like to add renewed thanks to Charles Parsons, and also to Michael Friedman, Steven Davis, Stefano Predelli, and Brian Epstein for helpful comments on this introduction. A special expression of appreciation for Robert Nozick, who invited me to publish my dissertation in this series. He inspired and he challenged, and his death is a great loss for his friends and for philosophy.

Dagfinn Føllesdal

Chapter One

Referential and Extensional Opacity

1. Introduction

Referential opacity, the phenomenon that substitutivity of identity and other logical principles relating to singular terms break down in certain contexts, might seem to be a rather incidental feature of our language. Apparently, little attention was paid to it by ancient and medieval logicians and philosophers. The Megaric logician Eubulides of Miletus (fourth century B.C.), who is best know for inventing the paradox of the Liar, is reported to have been concerned about Electra, who knows and does not know her veiled brother.[1] And in the fourteenth century a group of Oxford logicians, notable William Heytesbury and his pupil Billingham, were concerned with what we would now call the difference between the notional and relational sense of certain verbs which relate to mental activity, like 'knows that' and 'promises that'.[2]

---

[1] Eubulides was a pupil of Euclid of Megara, who in turn was a pupil of Socrates and the founder of the Megaric school. Eubulides' concern with Electra is reported in Diogenes Laertius, *De clarorum philosophorum vitis, dogmatibus et apophtegmatibus*, ed. C.G. Cobet (Paris, 1888), II, p.108. (The reference is from Bocheński, *Ancient Formal Logic*, p.100)

[2] This was recently discovered by a Harvard student, Mr. Waud H. Kracke, who found that an unedited manuscript by Billingham (Bibliothèque Nationale, Cod. Lat. 14, 715) contains a discussion of the relational vs. the notational sense of

Starting with Frege, however, referential opacity has become increasingly important by giving impetus to numerous theories of meaning. And it is still one of the main stumbling blocks for every theory of meaning. Thus in *Meaning and Necessity*, which might be considered an attempt to overcome the logical and semantical problems raised by referential opacity, or as Carnap calls it "the antinomy of the name-relation," Carnap lists five earlier attempts to overcome these difficulties, all of which he finds unsatisfactory, namely the methods of (1) Frege, (2) Quine, (3) Church, (4) Russell and (5) attempts to get by with a purely extensional language.[3]

One conspicuous reason why so much effort has been spent trying to overcome these difficulties, is that a proper understanding of the identity relation, with which the phenomenon of referential opacity is intimately connected, is important both in logic and philosophy. Another reason, which has perhaps weighed even more heavily, is that the contexts which are referentially opaque all seem to relate to phenomena and problems which traditionally have been considered to be of particular philosophical importance; necessity and possibility, laws of nature, moral obligation, knowledge and belief, love and hate, hope and fear, and apparently human acts and attitudes in general.

In this thesis, we will concentrate on the contexts which seem to be most readily accessible for logical analysis, viz., the logical modalities. These contexts and their semantic properties will be investigated in Chapters Two–Six. In Chapter Seven an attempt will be made to carry some of these results over to other types of opaque contexts.

---

certain verbs, together with a proposed syntactical criterion by which writes could make clear which of the two senses is intended. An English translation of the manuscript will be included in Mr. Kracke's B.A. thesis (Harvard, spring 1961).

[3] A lucid survey of these five methods is given in §32 of *Meaning and Necessity*.

2. *Criteria for Referential and Extensional Opacity*[4]

Criteria for referential opacity are most easily formulated by the help of the notion of a referential position, or occurrence, of a (singular) term.

When logical reasoning is formalized by the help of quantification theory, three kinds of expressions get prominence: singular terms, general terms, and sentences. In the quantificational schemata they have their counterparts in the variables, predicate letters, and sentence letters respectively.

Correspondingly, we get three criteria for *referential position* of singular terms:

1'.a   Referential position of a singular term in a singular term: interchangeability of co-referential terms *salva designatione.*

1'.b   Referential position of a singular term in a general term: interchangeability of co-referential terms *salva extensione.*

1'.c   Referential position of a singular term in a (closed) sentence: interchangeability of co-referential terms *salva veritate.*

Along with referential position of singular terms, we might talk about *extensional position* of general terms:

2'.a   Extensional position of a general term in a singular term: interchangeability of co-extensive terms *salva designatione.*

2'.b   Extensional position of a general term in a general term: interchangeability of co-extensive terms *salva extensione.*

2'.c   Extensional position of a general term in a (closed) sentence: interchangeability of co-extensive terms *salva veritate.*

Likewise we might talk about *truth functional position* of sentences:

---

[4] All the criteria and definitions listed in this section are due to Professor Quine. They were presented in his course *Philosophy 148, Philosophy of Language* in the spring of 1960. Most of them are also found in Quine's recent book *Word and Object.*

3'.a Truth functional position of a sentence in a singular term: interchangeability of sentences with the same truth-value *salva designatione.*

3'.b Truth functional position of a sentence in a general term: interchangeability of sentences with the same truth-value *salva extensione.*

3'.c Truth functional position of a sentence in a sentence: interchangeability of sentences with the same truth-value *salva veritate.*

Open sentences can presumably be assimilated with general terms in the above definitions, both kinds of expressions having extensions. Comparison of the open sentences 'x is warmer than y' and '(Ex) (x is warmer than y)' which have *extensions,* with the closed sentence '(Ex) (Ey) (x is warmer than y)' makes it natural to say that a truth value is the extension of a closed sentence. We are thereby led to assimilate general terms and open sentences with closed sentences, saying that they all have extensions.

The nine criteria above can then be reduced to four:

1'.a remains unchanged:

Def. 1.a Referential position of a singular term in a singular term: interchangeability of co-referential terms *salva designatione.*

1'.b and 1'.c collapse into:

Def. 1.b Referential position of a singular term in a general term or sentence: interchangeability of co-referential terms *salva extensione.*

2 '.a and 3 '.a collapse into:

Def. 2.a Extensional position of a general term or sentence in a singular term: interchangeability of co-extensional expressions *salva designatione.*

2 '.b, 2 '.c, 3 '.b and 3 '.c collapse into:

Def. 2.b Extensional position of a general term or sentence in a general term or sentence: interchangeability of co-extensional expressions *salva extensione.*

Such an assimilation of general terms to open and closed sentences might, of course, be questioned, since it might turn out to be misleading. We shall, however, find that general terms and open and closed sentences behave alike as far as our purposes are concerned, and that their behavior differs from that of singular terms.

With these four definitions at hand we may now define *referential transparency and opacity,* and parallel to it, *extensional transparency and opacity*:

> Def. 3 A construction, or mode of containment, φ, is {referentially/extensionally} transparent if and only if for every expression which may be an ingredient in φ, every position which is {referential/extensional} in the expression is {referential/extensional} in the product of the construction.

An expression ψ may be an ingredient in a construction φ if and only if φ (ψ) is well formed.

Since obviously every singular term is in a referential position with respect to itself, and every general term or sentence in an extensional position with respect to itself, we have by Def. 3:

> (i) any {singular term/general term or sentence} which is an ingredient of a {referential/extensional} construction, has a {referential/extensional} position in the product of the construction.

Constructions which are not {referentially/extensionally} transparent will be called {referentially/extensionally} opaque. Instead of speaking of an extensionally transparent construction we will for short speak of an *extensional construction*. An extensionally opaque construction will similarly for short be called a *non-extensional construction*.

3. Alternative Criteria for Referential Opacity

The criteria for referential and extensional position given above go back to Frege.[5] Treating general terms and sentences as referring to classes and truth values, Frege did not have to distinguish four criteria as we did in the preceding section (1.a, 1.b, 2.a and 2.b), but got by with just one: interchange of co-referential expressions.

An independent criterion for referential position was proposed by Quine in "Notes on existence and necessity" (1943), p.82. According to Quine,

> a position of a singular term is referential in a context only if *existential generalization* with respect to the position is permitted.

Similarly, there can, according to Quine, be no pronominal cross reference into modal contexts, or more precisely:

> any occurrence of indefinite singular terms and the occurrences of its associated pronouns must all have referential position with respect to some *one* containing sentence.[6]

For the variables of quantification theory this means that,

> all occurrences of variables bound by the same quantifier, except the occurrence in the quantifier itself, must have referential position in the open sentence which is the scope of the quantifier.

One might wonder whether the three criteria for referential occurrence—substitutivity of identity, existential generalization, and pronominal cross reference—are equivalent. Thus it has been argued that we may quantify into modal contexts and perform existential generalization with respect to singular terms occurring in them, but that substitutivity of identity breaks down in such

---

[5] Frege, "Über Sinn und Bedeutung" (1892), esp. pp.64 and 76 of Black's translation in *Philosophical Writings of Gottlob Frege*.

[6] This principle was presented by Quine in his lectures in the Spring of 1960. Cf. also *Word and Object*, pp.147-148.

contexts. This would indicate that the criterion of substitutivity of identity is stricter than the other two criteria.

In Chapter Three of this thesis we shall, however, find that the interpretation of the quantifiers in modal contexts apparently requires that identity by substitutive in these contexts. Therefore, there seems to by no evidence that the criteria are non-equivalent.

4. Examples and Further Characteristics of Referential and Extensional Opacity

Examples of constructions which are referentially and extensionally transparent are provided by truth functional logic and quantification theory. For if an expression is in {referential/extensional} position in an ingredient of a truth-functional or quantificational construction, then it is in {referential/extensional} position in the product of the construction.

The most conspicuous example of constructions which are referentially and extensionally opaque are quotations. Due to the referential opacity of quotations, quotations should not be regarded as *contexts* at all if one wants to retain the principle that identity is universally substitutive, i.e., substitutive in all contexts.

As Quine has pointed out,[7] quotations may be used to illustrate that,

> a position which is {referential/extensional} within one context, may be {non-referential/non-extensional} within a broader context, and yet {referential/extensional} within a still broader context.

Thus the positions of 'Tully' in

> 'Tully' refers to a Roman
> 'Tully denounced x' is true of Catiline
> and   'Tully was a Roman' is true

are referential, although the positions are non-referential in the quotation "contexts" 'Tully', 'Tully denounced x', and 'Tully was a Roman'. So the

---

[7] *From a Logical Point of View*, p.141; *Word and Object*, p.146.

constructions 'refers to', 'is true of' and 'is true' cancel quotation marks, so to say. Other expressions of the same type are 'is co-referential with' and 'is co-extensional with', in short the key expressions of the so-called theory of reference. Another group of expressions, the key-expressions of the so-called theory of meaning, viz., 'is analytic', 'is synonymous with', etc., have the property that, although they do not cancel quotation marks, they make quotations behave like modal contexts; interchange of analytic equivalents and of synonyms may take place within the quoted expression without any change of the reference or truth value of the product of the construction. Examples are:

> 'My bachelor brother' is synonymous with 'My adult unmarried brother'
> 'x is a bachelor' is analytically equivalent to 'x is an unmarried man'
> 'No bachelor is married' is analytic.

5.   Interrelations Between Referential Opacity and Extensional Opacity

One may prove that *every extensional construction is referentially transparent*, on the hypothesis that

> H   a singular term μ cannot be in a referential position in another
> singular term μ" without being in a referential position in a
> general term or sentence μ' which in turn is in an extensional
> position in the singular term μ".

The plausibility of this hypothesis will be discussed below. But let us first give the proof:[8]

Let μ be a singular term, and ⌐ . . μ . .⌐ an ingredient of a construction φ, and let

---

[8] The proof is due to Quine, who outlined it in his lectures in the spring of 1960. Hypothesis H and the discussion of it and a parallel hypothesis for general terms and sentences is original with this thesis.

1)   φ be extensional

Let further

2)   μ = ν

and

3)   the position of μ in ⌜. . μ . .⌝ be referentially transparent.

Then, by 2) and 3):

4)   the expressions ⌜. . μ . .⌝ and ⌜. . ν . .⌝ are (Case 1) co-extensional
     (if they are general terms or sentences) or (Case 2) co-referential
     (if they are singular terms)

Case 1. Since, by (i) of 2. and 1), ⌜. . μ . .⌝ is extensional in ⌜ φ(. . μ . .)⌝,
we may by 1) and 4) conclude that ⌜ φ(. . μ . .)⌝ and ⌜ φ(. . ν . .)⌝ are co-
extensive or co-referential. Hence, the position of μ in ⌜ φ(. . μ . .)⌝ is referential,
and the construction is referentially transparent.

Case 2. By 3) and hypothesis H, the position of μ is referential in some
general term or sentence μ' whose position is extensional in ⌜. . μ . .⌝. That
means firstly that the expression got from μ' by putting ν for μ is co-extensional
with μ', and secondly, by H and 1), that ⌜ φ(. . μ . .)⌝ and ⌜ φ(. . ν . .)⌝ are co-
extensional or co-referential. So the position of μ in ⌜ φ(. . μ . .)⌝ is referential,
and the construction is referentially transparent.

One might wonder how plausible hypothesis H is. It is hard to find
arguments for or against it. All I can report is that I am unable to find any
counter-example. There are certainly examples in which a singular term μ
occurs in another singular term μ'' without occurring in a general term or
sentence which occurs in μ''. Quotation contexts may serve as examples: the
singular term 'Socrates' occurs in the singular term ''Socrates'' without
occurring in any general term. But the important point is that 'Socrates' does not
occur in a referential position in ''Socrates''. So our hypothesis H is not
contradicted.

Hypothesis H and the above proof indicate, however, a way one might go if
one wanted to find an example or construct a system in which constructions
could be extensional without being referentially transparent: one might try to

find an example or permit some construction in which a singular term μ is in a referential position in another singular term μ" without being in a referential position in a general term or sentence which is in an extensional position in μ".

Parallel to H, one might propose the following hypothesis for general terms and sentences:

> A general term or sentence μ cannot be in an extensional position in another general term or sentence μ" without being in an extensional position in a singular term μ' which in turn is in a referential position in the general term or sentence μ".

On this hypothesis one could prove, by reasoning parallel to that of the proceeding proof, that every referentially transparent sentence is extensional. But the above hypothesis is plainly false. Already the truth functions provide examples of sentences which occur in extensional position in another sentence without occurring in an extensional position in any singular term of the kind mentioned in the hypothesis. Likewise, an open sentence is in an extensional position in the sentence got by prefixing a quantifier to it, but is not in an extensional position in any singular term of the type described.

These counter-examples to the hypothesis do, of course, not show that there are referentially transparent constructions which are not extensional. One might hope to prove that there are no such constructions along lines which are different from those we followed when we proved the converse.[9] In this thesis, we shall, however, see that there actually are constructions which are referentially transparent without being extensional. These are the constructions of quantified modal logic. In Chapters Three to Five we shall see that they have to be referentially transparent in order that we shall be able to quantify into them, and nevertheless non-extensional in order that modal distinctions shall not collapse.

---

[9] Quine, in *Word and Object*, notes that "if a construction is [referentially] transparent and allows substitutivity of concretion, it is extensional" (p.151, n.4). The results of this thesis accord well with this observation, since as we shall see (Theorem XVII), on order to avoid a collapse of modal distinctions in quantified modal logic, we have to disallow substitutivity of concretion.

Chapter Two

The Logical Modalities

6. Different Kinds of Modalities

Among the many types of expression in our language which are used to qualify something, some have attracted more attention than others and become the subject of formal investigations. Most prominent among these are the qualifications 'necessarily', 'possibly', 'contingently', the so-called alethic modalities, which many philosophers since antiquity have felt to be of particular philosophical importance. Among other modalities whose formal properties have been studied are the epistemic ones ('It is known that . . .', 'It is not known that . . .', etc.), some moral modes, e.g., the modes of obligation, of the deontic modalities ('It is obligatory that . . .', 'It is forbidden that . . .', etc.).[1] Many formal features recur from one type of these modalities to another. Professor G. H. von Wright,[2] among others, has pointed out that many of these features are found also in quantification theory. Thus, e.g., 'It is necessary that . . .' behaves in many ways like the universal quantifier, and 'It is possible that . . .' like the existential quantifier. This has led von Wright to consider quantifications "existential modalities."

---

[1] For a survey of these and other extensions of the notion of modality, see A.N. Prior, *Formal Logic*, pp. 215-220.

[2] In *An Essay in Modal Logic*, esp., pp. 2 and 19.

There is, indeed, a striking parallelism between quantifiers and modal operators. Writing '□' for 'necessarily', and '◇' for 'possibly', the following schemata are all valid (reference by page numbers are to von Wright's *An Essay in Modal Logic*):

<table>
<tr><td>Quantification theory:</td><td>Alethic modalities:</td></tr>
<tr><td>~(x)~Fx ≡ (Ex)Fx</td><td>~□~p ≡ ◇p</td></tr>
<tr><td>(x) Fx ⊃ Fx</td><td>□p ⊃ p</td></tr>
<tr><td>Fx ⊃ (Ex)Fx</td><td>p ⊃ ◇p
(von Wright's *Axiom of Possibility* [p.84])</td></tr>
<tr><td>(x) (Fx.Gx) ≡ . (x)Fx . (x) Gx
(Ex) (Fx v Gx) ≡ . (Ex) Fx v (Ex) Gx</td><td>□ (pq) ≡ . □p. □q
◇ (pvq) ≡ . ◇p v ◇q
(von Wright's *Axiom of Distribution* [p. 84], cf. also his *Principle of M-Distribution* [p. 12])</td></tr>
<tr><td>(x) (Fx ≡ Gx) ⊃ . (Ex)Fx ≡ (Ex)Gx</td><td>□ (p≡q) ⊃ . ◇p ≡ ◇q
(von Wright's *Principle of M-Extensionality* [p.12])</td></tr>
</table>

The treatment of the "existential modalities" is, according to von Wright, sometimes called quantification theory and is usually not regarded as a branch of modal logic.[3]

As we saw in section 4, quantificational constructions are referentially and extensionally transparent. Let ⌜(α)φ⌝ be a universally quantified expression. If in the expression φ which follows the quantifier we substitute for a singular term a co-referential term, for a general term a co-extensional term, or for a sentence a sentence with the same truth-value, then if we do not violate the restrictions on substitution, we get a sentence with the same truth-value as the original one. The parallelism between modal operators and quantifiers should then lead us to expect that also modal constructions are referentially and extensionally transparent. However, they are not. For let φ be a statement which is necessarily true, and ψ a statement which is true, but not necessarily true. Then if in

---

[3] *An Essay in Modal Logic*, p.2.

(1) $\ulcorner \Box \varphi \urcorner$

we substitute ψ for φ we get

(2) $\ulcorner \Box \psi \urcorner$

which, unlike (1), is false.

Thus, at least on one point, we have found an important discrepancy between the modal operators and the quantifiers. This makes us suspect that the parallelism is not quite as close as a first comparison might lead us to think. Von Wright does not treat the "existential modalities" in his *An Essay in Modal Logic*, but refers to his treatment of uniform quantification theory in *On the Idea of Logical Truth*.[4] And indeed, by comparing his treatment of uniform quantification theory there with his treatment of the alethic modalities in *An Essay in Modal Logic*, it is easily seen that the true necessity sentences correspond, not to the *true* universally quantified sentences, but rather to the *logically true* universally quantified sentences, i.e., to those of them which exemplify valid quantificational schemata.

Now, as is well known, a universally quantified schema in uniform quantification theory is valid if and only if the inside structure exhibits the form of a valid truth functional schema. Rather therefore, than comparing e.g., the operator '□' with '(x)', one should compare '□' with the words 'is logically true' or 'is valid'.

Since validity, non-validity, consistency and inconsistency are retained under the operation of interchange of equivalents, the parallel between '□' and 'is valid' teaches us that we may expect that interchange of logically equivalent expressions will leave the truth value of modal sentences undisturbed.

The expressions 'is logically true' and 'is valid' attach to names of sentences, and as pointed out by Quine,[5] since quotations are opaque, the parallel between the expressions 'is logically true' and 'is valid' and the logical operators may serve to remind us of the opacity of modal contexts. Since, as just noticed, interchange of equivalents can be expected to work in modal contexts, the opacity of modal contexts may, however, be expected to be dissimilar to the opacity of quotation contexts.

---

[4] *On the Idea of Logical Truth* I-II. Essay I is reprinted in von Wright's *Logical Studies*, pp. 22-43.

[5] In "Three grades of modal involvement," p.74.

In this thesis, the opacity of modal contexts will be examined. Attention will be focused on the logical modalities. The results will be carried over to the other modalities in Chapter Seven.

7. Unquantified Modal Logic

The alethic modes can all be reduced to that of necessity, or alternatively, to that of possibility. Possible is that whose negation is not necessary, necessary is that whose negation in not possible. Customarily, the idea of necessity is expressed in modal logic by an operator '□' which attaches to sentences to form new sentences. In this section, attention will be paid solely to its use as a statement operator, i.e., an operator applying to *closed* sentences.

Obviously, unless modal distinctions shall collapse, contexts governed by '□' have to be extensionally opaque. For if they were not, then in

(1)   □ (9>7)

we could replace '9>7' by any true statement, and hence prove that any statement, if true at all, is necessarily true. Likewise we could prove that any false statement would be necessarily false.

This observation applies regardless of how one decides to interpret the word 'necessity', whether one has in mind some notion of natural or physical necessity, or, as in this and the following sections, some notion of alethic or logical necessity, according to which □ (. . .) if and only if ' . . . ' is logically, or analytically, true. The fact that necessity contexts thus are bound to be extensionally opaque, regardless of how 'necessity' is conceived, as long as we don't want modal distinctions to collapse, may seem disastrous for any attempt to analyze contrary-to-fact conditionals or disposition terms by appeal to some kind of natural necessity. For if every extensionally opaque construction should turn out to be referentially opaque, and if the three criteria for referential transparency mentioned in sections 2 and 3 are equivalent, this would mean that one could not quantify into natural necessity-contexts.[6]

---

[6] Quine, in his "Notes on existence and necessity" shows that necessity contexts are opaque and continues: "These observations apply, naturally, to the prefix

In Chapter Three we shall see that the three criteria for referential opacity are apparently equivalent. And it does also seem as if the modal contexts are *referentially* opaque. For if in (1) we substitute for '9' 'the number of planets', we get

(2) $\Box$(the number of planets > 7)

which is presumably false.[7]

'necessarily' only in the explained sense of analytic necessity; and correspondingly for possibility, impossibility, and the necessary conditional. As for other notions of necessity, possibility, etc., for example, notions of physical necessity and possibility, the first problem would be to formulate the notions clearly and exactly. Afterwards we could investigate whether such notions involve nondesignative occurrences of names and hence resist the introduction of pronouns and exterior quantifiers." (p.124).

If such notions of necessity etc., can be found, then they would supply an example to the effect that there are constructions which are extensionally opaque, and nevertheless referentially transparent. In Chapters Three through Five we shall find that, when certain conditions are satisfied, already the notion of logical necessity provides examples to this effect.

[7] That the principle of substitutivity of identity breaks down in modal contexts was observed by Quine in "Whitehead and the rise of modern logic" (1941), pp. 141-142, esp. note 26, and p. 148.

In his review of this essay in *Journal of Symbolic Logic*, 7 (1942), pp. 100-101, Alonzo Church pointed out that Quine's example ('$\Diamond$ (the number of planets < 7)' is true, while '$\Diamond$ (9<7)' is false) involves the description or class abstraction 'the number of planets' which both in *Principia Mathematica* and in Quine's own *Mathematical Logic* would be construed contextually: " . . . any formal deduction must refer to the unabbreviated forms of the sentences in question, and the unabbreviated form of the first sentence is found to contain no name of the number 9." (p.101).

Church adds that he would prefer a system in which class abstract and description are construed as names, and hence are not contextually defined. In a system of this kind, Quine's argument shows that "a non-truth-functional operator, such as $\Diamond$, if it is admitted, must be prefixed to names of propositions rather than to sentences." (p.101).

These two remarks by Church have been followed up by Smullyan, who, first in his review of Quine's "The problem of interpreting modal logic" and thereafter in "Modality and description," has tried to show that "the logical modalities need not involve paradox when they are referred to a system in which descriptions and class abstracts are contextually defined." ("Modality and description," p.35). Smullyan says that both when the modal operators are construed as statement operators *and* when they are construed as *sentence* operators (p.34), "the modal paradoxes arise out of neglect of the circumstance that in modal contexts the scope of incomplete symbols, such as abstracts or descriptions, affect the truth value of those contexts." (p.37).

Both Miss Barcan, in her review of "Modality and description," and Fitch, in his article "The problem of the Morning Star and the Evening Star," have expressed that they are largely in agreement with Smullyan, although they have pointed out some slips and proposed some minor revisions in Smullyan's paper.

Quine, commenting on Smullyan's proposal in *From a Logical Point of View*, remarks, however, that " . . . referential opacity remains to be reckoned with even when descriptions and other singular terms are eliminated altogether." (p.156).

And Neil Wilson, in *The Concept of Language*, has argued that even when applied to descriptions and singular terms Smullyan's technique has a decisive shortcoming: it works for some examples, but apparently not for all. Thus, writing '$(\imath x) (Mx)$' for 'the author of *Marmion*', '$(\imath x) (Wx)$' for 'the author of *Waverly*', and 'S' for 'is Scotch', we can from two true premises deduce a false conclusion:

$$(\imath x) (Wx) = (\imath x) (Mx)$$
$$\Box([(\imath x) (Wx)] \, S(\imath x) (Wx) \equiv [(\imath x) (Wx)] \, S(\imath x) (Wx))$$
$$\Box([(\imath x) (Wx)] \, S(\imath x) (Wx) \equiv [(\imath x) (Mx)] \, S(\imath x) (Mx))$$

"Here it is not a question of substituting a description for a name and then judiciously choosing the scope. If we wish to substitute '$(\imath x) (Mx)$' on the RHS of '$\equiv$' in the second premise, the scope is already selected. Thus we are forced to conclude that '$(\imath x) (Wx)$' and '$(\imath x) (Mx)$' are not interchangeable in all modal contexts and that within a modal language the author of *Waverly* is not identical with the author of *Marmion*. Since identity is a transitive relation we dare not say that nevertheless they are both identical with Walter Scott. In general then, the individuals of a modal language are individual concepts rather than ordinary concrete individuals."

We will, however, return to this apparent breakdown of the substitutivity of identity in sections 17 and 18, and to natural necessities and contrary-to-fact conditionals in section 21.

8. *Quantified Modal Logic*

If substitutivity of identity breaks down in modal contexts, and if all the criteria for referential opacity listed in sections 2 and 3 are equivalent, we should expect that also the criterion of existential generalization would break down when applied to modal contexts. And also that it should be impossible to quantify into modal contexts, i.e., impossible to treat the modal operator '□' as a *sentence* operator, applicable to all kinds of sentences, *open* or closed, to form new sentences.

And, certainly, there is a curious air about e.g.,

     (1)   (Ex) □ (x>7)

Quine has asked: " . . . would 9, that is, the number of planets, be one of the numbers necessarily greater than 7?"[8] and pointed out that such an affirmation would be true in the form

     (2)   □ (9>7)

and false in the form

     (3)   □ (the number of planets >7)

Nevertheless, logicians have tried to construct systems of quantified modal logic. And for good reasons. For, as Carnap has put it: "Any system of modal logic without quantification is of interest only as a basis for a wider system

---

Smullyan's proposal and Quine's and Wilson's criticism of it will be commented upon in section 18.

[8] In "Notes on existence and necessity," pp. 123-124. See also *From a Logical Point of View*, pp. 147-148.

including quantification. If such a wider system were found to be impossible, logicians would probably abandon modal logic entirely."[9]

Hintikka has made the need for a quantified modal logic evident by showing how in the case of deontic logic introduction of quantifiers must be regarded not only as a means of making current systems more comprehensive, but as, "indispensable for any satisfactory analysis of the notions with which *every* system of deontic logic is likely to be concerned."[10]

By showing that central notions in ordinary language like those of obligation, forbiddance, permission, and commitment, call for an analysis in terms of quantifiers, Hintikka has, it seems, not only made a good case for the *need* for a quantified modal logic, but also, unless our ordinary reasoning and discourse in this area is extraordinarily muddled and confused, for the *possibility* of one. Added support for the view that quantified modal logic is possible is rendered by the fact that actual systems of quantified modal logic have been constructed and claimed to work.[11] But the difficulties remain. One, the difficulty of making sense of existential generalization, has already been mentioned. Many others have been pointed out by critics of modal logic, notably by Quine.

---

[9] *Meaning and Necessity*, p. 196.

[10] Hintikka, "Quantifiers in deontic logic," p.3

[11] In "Modalities and quantification" (1946) Carnap gives a restricted completeness proof for his extension of S5 (T 12-3, p. 63 of Carnap's article. See also p.53 of his article). In his review of this article in *Journal of Symbolic Logic* Bernays gives his reasons for believing that a full completeness proof for Carnap's system can probably be given.

In 1951, H. Rasiowa in "Algebraic treatment of the functional calculi of Heyting and Lewis" extended the Lewis calculus S4 into a quantified modal logic and proved the extension complete. Kripke in "A completeness theorem in modal logic" (1959) extended S5 into a quantified modal logic with identity and proved the extension consistent and complete. Independently of Kripke, A. Bayart, in "La correction de la logique modale de premier et second ordre S5" (1958) extended S5 into a first order and a second order functional calculus and proved both consistent. In "Quasi-adéquation de la logique de second ordre S5 et adéquation de la logique modale de premier ordre S5" (1959), Bayart proved the completeness of his first order modal calculus, and the completeness (in the sense of Henkin's completeness theorem for the non-modal functional calculus of second order) of his second order modal calculus.

Let us now first survey the systems of quantified modal logic constructed till now (section 9), and then in the following sections review the difficulties confronting a quantified modal logic and see whether and how they may be overcome.

9. *Systems of Quantified Modal Logic*

1. The first system of quantified modal logic to be constructed seems to be the extension of the Lewis calculus S2[12] presented in Miss Ruth C. Barcan (Mrs. Ruth Barcan Marcus), "A functional calculus of first order based on strict implication" (1946).

Miss Barcan's system is got from S2 by adding quantifiers and a standard system of axiom for quantification theory, with, in effect, an '□' prefixed to each axiom,[13] together with the special axiom

(1) $\ulcorner \Diamond(E\alpha)\varphi \prec (E\alpha)\Diamond\varphi \urcorner$

Some further results relating to this system and to a similar extension of the Lewis calculus S4 were later presented by Miss Barcan in "The deduction theorem in a functional calculus of first order based on strict implication" (1946), and "The identity of individuals in a strict functional calculus of second order" (1947).

2. The next system to be published was Carnap's system MFC, an extension of a simplified version of S5 (due to M. Wajsberg [1933]), in "Modalities and quantification" (1946). As remarked by Carnap (middle of p.33) no forms of quantified modal logic had been published when Carnap submitted his paper. A similar system, S_2 (not to be confused with the Lewis system S2), is

---

[12] Postulate sets for all the systems of sentential modal logic mentioned in the thesis are given in Appendix I.

[13] All of Miss Barcan's axioms are of the form $\ulcorner \varphi \prec \psi \urcorner$, $\ulcorner \varphi \equiv \psi \urcorner$, or $\ulcorner (\alpha_1)(\alpha_2) \ldots (\alpha_m)(\varphi \prec \psi) \urcorner$. From this it follows, by theorem 18.7 in Lewis and Langford, Barcan's theorem 39 and substitution that if φ is an axiom, then $\vdash \ulcorner \varphi \equiv \Box\chi \urcorner$. This is remarked by Miss Barcan on p.118 of her paper "The deduction theorem in a functional calculus of first order based on strict implication" (1946).

given in Carnap's *Meaning and Necessity* (1947, 1956), §41; see also §§40 and 43.

A main difference between MFC and S_2 is that S_2 contains definite descriptions. A characteristic feature of Carnap's systems is that the values of his individual variables are not individuals, but individual concepts.[14]

3. In 1948, in "Intuitionistic modal logic with quantifiers," Fitch extended the Heyting intuitionistic calculus to modal concepts.

4. Miss H. Rasiowa, in her "Algebraic treatment of the functional calculi of Heyting and Lewis" (written 1950, published in *Fundamenta Mathematicae* for 1951, which appeared in 1953), constructed a system of quantified modal logic based upon the Lewis calculus S4. This system is discussed also in Rasiowa and Sikorski's two joint papers "Algebraic treatment of the notion of satisfiability" (1953) and "On existential theorems in non-classical functional calculi" (1954).

Leon Henkin in his review of Miss Rasiowa's paper in the *Journal of Symbolic Logic* (1953) regrets that "the relationship between the modal calculus considered by Barcan (XI, 96) ["A functional calculus of first order based on strict implication"] and Carnap (XIII, 218) ["Modalities and quantification"] is not made clear" (p.73). Here are some main differences: (1) While Miss Rasiowa's system is an extension of S4, Carnap's is an extension of S5.[15] (2) Miss Barcan's main system is an extension of S2. Her extension of S4 might be expected to coincide with Rasiowa's calculus. Miss Barcan's system seems, however, to be *stronger*, since she has included among her axiom schemata one special schema for the mixture of quantifiers and modal operators, viz., her axiom schema 11 (p.2):

$$(1) \qquad \ulcorner \Diamond(E\alpha)\varphi \prec (E\alpha)\Diamond\varphi \urcorner$$

Miss Rasiowa has no special axiom for the mixture, neither does she prove any corresponding theorem. And it seems unlikely that schema (1) can be

---

[14] "Modality and quantification," pp.37-38 and p.64, *Meaning and Necessity*, §§10, 40 (esp. top of p.180), 43, and 44.

[15] That Carnap's system is an extension of S5 is made clear in note 8 of Carnap's paper.

proved in her system, since "Brouwer's axiom" $\ulcorner\varphi\supset\Box\Diamond\varphi\urcorner$, which is independent of S4 seems to be needed in order to prove it.[16]

Another, less important difference between Miss Rasiowa's and Miss Barcan's extensions of S4 is that Miss Rasiowa, but not Miss Barcan, includes among her axioms the Lewis axiom B5 ($\ulcorner\varphi\prec\sim\sim\varphi\urcorner$) which as shown by J. C. C. McKinsey[17] is derivable from axioms B1, B2, B3, and B6. Miss Rasiowa also includes the Lewis axiom B8 ($\ulcorner\Diamond(\varphi.\ \psi)\prec\Diamond\varphi\urcorner$), which as pointed out already by Lewis[18] is derivable from the remaining axioms in S4. Miss Barcan, too, includes this axiom, but this is natural, since her quantified S4 is constructed as an extension of her quantified S2, in which the axiom B8 is not redundant. Like Miss Barcan, Miss Rasiowa prefixes (in effect) an '$\Box$' to all her axioms of the functional calculus. (Cf. note 13, this chapter).

5. In "The logic of causal propositions" (1951) Arthur W. Burks constructs a system of quantified logic containing operators both for logical and for physical, or natural necessity. As far as the logical modalities are concerned, his system seems to be Gödel's basic system[19] (which is equivalent to von Wright's system M and to Fey's system T,[20] stronger than S2 and weaker than S4) supplemented with quantifiers, a set of axioms for quantification theory, and the special axiom $\ulcorner(\alpha)\Box\varphi\supset\Box(\alpha)\varphi\urcorner$ for the mixture.

6. The same year, 1951, Church, in "A formulation of the logic of sense and denotation" outlined a system containing quantifiers and modal operators

---

[16] By adding "Brouwer's axiom" to the Gödel-Feys-von Wright system one gets a system weaker than S5 and incomparable in strength with S4. In section 15, formula (1) above will be proved in a quantified modal logic based on this "Brouwerian" system. A proof of (1) within a quantified modal logic based on S5 has been given by A. N. Prior in "Modality and quantification in S5" (1956), the result was stated without proof already by Carnap in his extension of S5 in *Meaning and Necessity* (1947) (p.186, theorem m).

[17] "A reduction in the number of postulates for C. I. Lewis' system of strict implication" (1934).

[18] *Symbolic Logic*, p.501.

[19] The system is in Gödel's paper "Eine Interpretation des intuitionistischen Aussagenkalküls" (1933) without "Becker's axiom" ($\ulcorner\Box\varphi\supset\Box\Box\varphi\urcorner$), i.e., the system consisting of the two axioms: $\ulcorner\Box\varphi\supset\varphi\urcorner$ and $\ulcorner\Box\varphi\supset.\ \Box(\varphi\supset\psi)\supset\Box\psi\urcorner$ and the rule of inference: if $\vdash\varphi$, then $\vdash\ulcorner\Box\varphi\urcorner$.

[20] That Gödel's basic system is equivalent to M and to T is proved by Sobocinski in his "Note on a modal system of Feys-von Wright" (1953).

which differs from the other systems so far constructed by being based on Frege's theory of meaning.[21]

In Church's system modal operators attach not to sentences, i.e., names of truth-values, but to names of propositions. For this reason, his system is both in formal and semantical respects radically different from all other systems of modal logic constructed so far. Church's system will be discussed in Appendix II. In all other sections of this dissertation, when systems of modal logic are discussed and results about them formulated, only systems in which the modal operators attach to *sentences*, open or closed, will be called 'systems of modal logic'. This convention will save us from a number of qualifications of the type: 'except systems in which the modal operators do not attach to sentences'.

7.   Fitch's system of quantified modal logic in *Symbolic Logic* (1952), pp.164-166, is essentially an extension of S4 without the principle of excluded middle and without Miss Barcan's special axiom 11 ((1) above: $\ulcorner\Diamond(E\alpha)\varphi \prec (E\alpha)\Diamond\varphi\urcorner$). Changes in the system resulting from adding the principle of excluded middle are discussed, likewise the consequences of putting a restriction on reiterations in a strict subordinate proof (which would change the system into S2), or of assuming axioms like $\ulcorner\Diamond(E\alpha)\varphi \supset (E\alpha)\Diamond\varphi\urcorner$ (Miss Barcan's axiom 11 with '$\supset$' instead of '$\prec$') or $\ulcorner(\alpha)\Box\varphi \supset \Box(\alpha)\varphi\urcorner$ (Burks's axiom).

8.   In 1958, A. Bayart, in "La correction de la logique modale de premier et second ordre S5," extended a system equivalent to the Lewis system S5, formulated by the help of Gentzen sequences, into a first and a second order functional calculus, adding no new axioms except the axioms of the non-modal first order and second order predicate calculi, and proved both extensions consistent. The next year, in "Quasi-adéquation de la logique modale de second ordre S5 et adéquation de la logique modale de premier ordre S5," Bayart proved both his first order and his second order calculi complete, adapting to his two extensions of S5 the general method of obtaining completeness proofs for functional calculi due to Henkin (1947, 1949).

9.   In 1959, in "A completeness theorem in modal logic," Kripke supplemented the Gödel system of axioms and rules for S5 given in Prior's

---

[21] Some main features of Church's system were presented in an address read at the eighth meeting of the Association for Symbolic Logic on February 23, 1946. An abstract of the address, with an addendum, dated April 29, 1946, was published in *The Journal of Symbolic Logic*, XI (1946), p.31.

Formal Logic with a standard formalization of the classical first-order predicate calculus with equality and proved the resulting system consistent and complete.

10-11. In *An Essay in Modal Logic* (1951) von Wright combined epistemic modalities with quantifiers. And in "Quantifiers and deontic logic" (1957), Hintikka combined deontic modalities with quantifiers.

As far as I know, these are all the systems of quantified modal logic which have been proposed till now. In his review in *The Journal of Symbolic Logic* (1951) of the second edition of Lewis and Langford's *Symbolic Logic* and again in "A formulation of the logic of sense and denotation" (1951), n.23, Church remarks that Chapter IX of Lewis and Langford "contains the earliest discussion of quantifiers in connection with strict implication."[22] The combination of modal operators and quantifiers considered by Lewis and Langford does, however, not qualify as a quantified modal logic in our sense of the term. Lewis and Langford never consider quantifying *into* modal contexts, but confine their attention to quantification *within* such contexts, i.e., their quantifiers are always situated to the right of the modal operators and have *no modal operators within their scope.* Lewis and Langford hence always treat the modal operators as *statement* operators, and their "quantified modal logic" does not raise the serious problems which we encounter trying to interpret the systems of quantified modal logic mentioned in this section, systems in which the modal operators are use as *sentence* operators.[23]

10. Difficulties Relating to Quantification into Modal Contexts

1.   The difficulty mention in section 8 concerning the sense of

(1) $(Ex)(x>7)$

---

[22] Church's review of Lewis and Langford's *Symbolic Logic*, p.225.

[23] The importance of this distinction is make clear in Quine's "Three grades of modal involvement" (1953) to which we will return in part 8 of the next section.

which Quine raised in his "Notes on existence and necessity" (1943), was, as far as I know, the first objection ever raised against quantification into modal contexts.[24]

2. In the same paper ("Notes on existence and necessity," 1943), Quine also points out that, as already remarked, substitutivity of identity breaks down in the context 'necessarily . . . ', at least when 'necessarily' is taken in the "analytic" sense (see section 6, above), and that this context, therefore, is similar to the context of single quotes, which does not admit pronouns which refer to quantifiers anterior to the context (p.123). Professor Alonzo Church, in his review of Quine's "Notes on existence and necessity" in *The Journal of Symbolic Logic*, VIII (1943), pp. 45-47, agreed with Quine that modal contexts are opaque. But Church argued that this does not prevent variables within the modal context from referring to a quantifier anterior to the context, *provided the quantifier has an intensional range*—a range, for instance, composed of attributes rather than classes.

3. In 1945-46, in letters to Carnap, quoted in Carnap's *Meaning and Necessity* (pp.196-197), Quine agrees that "adherence to an intensional ontology, with extrusion of extensional entities altogether from the range of values of the variables, is indeed an effective way of reconciling quantifications and modality." (p.197). But he points out that this is a more radical move than one might think, as becomes apparent when one tries to reformulate in intensional language the two statements:

(2)   The number of planets is a power of three
(3)   The wives of two of the directors are deaf

---

[24] The breakdown of the principle of *substitutivity of identity* in modal contexts was, however, noted by Quine already in "Whitehead and the rise of modern logic" (1941). (See note 7 above, this chapter.) Also, that same year, in his review of Russell's *Inquiry into meaning and Truth* (*The Journal of Symbolic Logic*, VI (1941), pp.29-30.), Quine called attention to the difficulties relating to quantification into belief contexts: "Moreover, he [Russell] never mentions the more difficult sort of contexts, wherein the matter following 'believes that' falls short of being a sentence because a variable in it is governed by a quantifier somewhere to the left of 'believes that'" (p.30).

It can be done, but the examples "give some hint of the unusual character which a development of it [an intensional language] adequate to general purposes would have to assume." (p.197).

4. In 1947, in "The problem of interpreting modal logic," Quine again took up the problem of making sense of existential quantifications containing modal operators. Let us suppose, he argued, that we try to make sense of such quantifications e.g., by the criterion:

> (i) An existential quantification holds if there is a constant whose substitution for the variable of quantification would render the open sentence true.

This criterion is only a partial (sufficient) one (because of unnamable objects), hence the 'if'. But it allows Quine to show that a quantified modal logic would have queer ontological consequences: "It leads us to hold that there are no concrete objects (men, planets, etc.) but rather that there are only, corresponding to each supposed concrete object, a multitude of distinguishable entities (perhaps "individual concepts," in Church's phrase). It leads us to hold, e.g., that there is no such ball of matter as the so-called planet Venus, but rather at least three distinct entities: Venus, Evening Star, and Morning Star" (p.47).

To show this, Quine uses "'C' for 'congruence' to express the relation which Venus, the Evening Star, and the Morning Star, e.g., bear to themselves and, according to empirical evidence, to one another. (It is the relation of *identity* according to materialistic astronomy, but let us not prejudge this.)." Then

Morning Star C Evening Star . $\square$(Morning Star C Morning Star)

Hence, by (i),

(4) (Ex) (x C Evening Star . $\square$(x C Morning Star))

But also,

Evening Star C Evening Star . ~$\square$(Evening Star C Morning Star)

so that, by (i),

(5)   (Ex) (x C Evening Star . ~□(x C Morning Star))

Since the open sentences quantified in (4) and (5) are mutual contraries, Quine concludes that there are at least two objects congruent to Evening Star. Similarly, by introducing the term 'Venus' a third such object can be inferred.

Parallel arguments may be used to show that the contemplated version of quantified modal logic is committed to an ontology repudiating classes and admitting only attributes, Quine adds. But the above argument concerning individuals is, of course, more disturbing for modal logicians.

5.   Later the same year, in his review of Ruth Barcan's "The identity of individuals in a strict functional calculus of second order," Quine remarks that Miss Barcan's system is "best understood by reconstruing the so-called individuals as 'individual concepts.' "[25]

6.   The article "Reference and modality" in Quine's *From a Logical Point of View* (1953) is a fusion of "Notes on existence and necessity" with "The problem of interpreting modal logic." But new arguments are added, notably arguments to the effect that we cannot properly quantify into a modal context.

First, lest the reader feel that the arguments against quantification into modal contexts always turn on an interplay between *singular terms* like 'Tully' and 'Cicero', '9' and 'the number of planets', 'Evening Star' and Morning Star', Quine re-argues the meaninglessness of quantification into modal contexts *without reverting to singular terms*. He points out that one and the same number x is uniquely determined by the conditions:

(6) $x = \sqrt{x} + \sqrt{x} + \sqrt{x} \neq \sqrt{x}$

and

(7)   There are exactly x planets.

Nevertheless, (6) has 'x>7' as a necessary consequence, while (7) has not. "Necessary greaterness than 7 makes no sense as applied to a *number* x; necessity attaches only to the connection between 'x>7' and the particular method (6), as opposed to (7), of specifying x." (p. 149).

Similarly, Quine argues,

---

[25] *The Journal of Symbolic Logic*, XII (1947), p.96.

(8)   (Ex)(necessarily if there is life on the Evening Star then there is life on x)

is meaningless, "because the sort of thing x which fulfills the condition:

(9)   If there is life on the Evening Star then there is life on x,

namely a physical object, can be uniquely determined by any of various conditions, not all of which have (9) as a necessary consequence. *Necessary* fulfillment of (9) makes no sense as applied to a physical object x; necessity attaches at best, only to the connection between (9) and one or another particular means of specifying x." (p.149).

7.   Suppose that, on order to overcome this difficulty, one were to retain within one's universe of discourse only objects x such that any two conditions uniquely determining x are analytically equivalent, i.e., such that

(ii) $(x)((y)(Fy\equiv. y=x). (y)(Gy\equiv. y=x). \supset \Box(y)(Fy\equiv Gy))$

or, if one permits quantification into modal contexts:

(ii') $(x)((y)(Fy\equiv. y=x) \supset \Box(y)(Fy \equiv. y=x))$

(ii), or (ii'), has, however, Quine points out, consequences which some modal logicians might be reluctant to accept, e.g.,

(10) $(x)(z)(x=z . \supset\Box(x=z))$

which is got by introducing the predicate '①=z' for 'F①' in the open sentence following '(x)' in (ii') and then closing and simplifying.

8.   In "Three grades of modal involvement" (1953), Quine adduces a different argument to the same effect, viz., that

(10) $(x)(y)(x=y . \supset.\Box(x=y))$

In any theory, whatever the shape of its symbols, an open sentence whose free variables are 'x' and 'y' is an expression of identity only in case it fulfills

(11) $(x)(y)(x=y . \supset. Fx\equiv Fy)$

in the role of 'x=y'. But, introducing the predicate '□(x=①' for 'F①' in (11) (or rather in 'x=y .⊃. Fx≡Fy' and then closing), as we can do if we admit quantification into modal contexts, we get, after simplification, (10) of the preceding argument over again:

$$(x)(y)(x=y \supset .\Box(x=y))$$

Quine adds that we do not have to infer from this that

$$\Box(\text{the number of planets} = 9)$$

if we accept some interference in the contextual definition of singular terms even when their objects exist, e.g., that we can't use them to instantiate universal quantification unless some special supporting lemma is at hand (p.80).

As a third consequence of quantification into modal contexts, in addition to (10) and the interference in the contextual definition of singular terms, we get Aristotelian essentialism, Quine points out. Not only can one have open sentences 'Fx' and 'Gx' such that

$$(Ex)(\Box Fx.Gx.\sim\Box Gx)$$

but one also must require that there are open sentences fulfilling

$$(x)(\Box Fx.Gx.\sim\Box Gx)$$

"An appropriate choice of 'Fx' is easy: 'x=x'. And an appropriate choice of 'Gx' is 'x=x.p' where in place of 'p' any statement is chosen which is true but not necessarily true." (p.81).

9. In *Word and Object* (1960), Quine draws a further, disastrous consequence of the requirement on open sentences to which he arrived in *From a Logical Point of View* (requirement (ii) above), viz., the consequence that every true sentence is necessarily true:

Let 'p' stand for any sentence. Introducing 'p .①=x' for 'F①' and '①=x' for 'G①' in (ii) (or rather in the open sentence following the '(x)' in (ii) and then closing), one gets:

$$(x)((y)(p.x=y \; .\equiv. \; y=x) \; . \; (y)(y=x \; .\equiv. \; y=x) \; .\supset \; \Box(y)(p.y=x \; . \equiv. \; y=x))$$

By universal instantiation one gets:

$$(y)(p.y=y \; .\equiv. \; y=y) \; . \; (y)(y=y \; .\equiv. \; y=y) \; . \supset \; \Box(y)(p.y=y \; . \equiv. \; y=y)$$

which reduces to

$$p \supset \Box p$$

Obviously, restricting the values of one's variables to intensional objects does not save one from this collapse of modal distinctions. So, unless quantification into modal contexts can be interpreted without assuming (ii), the prospects for a quantified modal logic are very gloomy indeed.

11. Interpreting Quantified Modal Logic

The notions of possible worlds (Leibniz), state descriptions (Carnap), or model sets of formulae (Hintikka) may be of help for seeing how modal logic fares when it is supplemented with quantifiers.

Starting from the simple, let us suppose that there are just two entities, a and b, and two possible universes, one of which is the actual one. Let us further suppose that four predicates, represented by the letters 'F', 'G', 'H', and 'K', are true or false of a and b in the two possible worlds as indicated below:

Possible non-actual world	Fa.~Ga.~Ha.Ka	~Fb.Gb.Hb.~Kb
Actual world	Fa.~Ga.Ha.~Ka	~Fb.Gb.~Hb.~Kb

Taking the necessary to be that which is true in all possible worlds, and the possible as that which is true in some possible world, we have not only:

$$\Box(Fa), \; \Box(Gb), \; \Box(\sim Ga), \text{ etc.}$$

but also

    (1)   (Ex) □(Fx)
    (2)   (Ex)◇(Kx)
            etc.

Obviously, the difficulties of making sense of existential generalizations like (1) and (2), and the other difficulties relating to quantification into modal contexts which were surveyed in the preceding section recur in connection with our example.

But before considering them, let us first eliminate another serious drawback of our example, relating to the notion of *iterated modalities*, like ⌜□□φ⌝, ⌜□(◇φv.ψ.□χ)⌝, etc. In order to make sense of them, our simple-minded notion of possible worlds will not suffice.

The iterated modalities may apparently be analyzed most satisfactorily if one conceives of possibility as a *relation*, and considers, not a set of worlds which are possible outright, but a set of worlds which are, or are not, *possible with respect to* another.[26]

Letting horizontal lines, dotted or dashed, represent possible worlds, writing 'XRY' for 'X is possible with respect to 'Y' and putting a vertical arrow from the line representing the possible world X to the line representing the possible world Y, if and only if XRY, iterated modalities might be represented as follows:

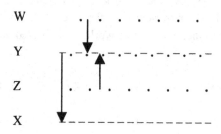

W

Y

Z

X

---

[26] This is the approach used by Kripke in "A completeness theorem in modal logic." I am indebted to this paper and to conversations with Kripke for the interpretation of iterated modalities outlined in this section.

The arrows in this diagram would indicate that the world Y is possible with repect to X, W to Y, and Z to Y.

We then would have the following rule for the reading and interpretation of iterated modalities. The outermost modal operator, if it is '◇' is read: 'There is some world Y, possible with respect to the actual one, such that in Y . . .' An operator '◇', which is within the scope of the outermost operator but of no other is read: 'Some world Z, possible with respect to Y, is such that in Z . . .' and so on. The reading of a necessity operator would correspondingly be: 'Any world Z, possible with respect to Y, is such that in Z . . . ', etc.

One basic law of the modalities is,

$$(3) \qquad \ulcorner \varphi \supset \Diamond \varphi \urcorner$$

which obviously just means that every world is possible with respect to itself, i.e., that the relation R (is possible with respect to) is (totally) *reflexive*:

$$(3') \qquad (X)\,(XRX)$$

The two most notable laws governing iterated modalities are the so-called "Becker axiom":

$$(4) \qquad \ulcorner \Diamond \Diamond \varphi \supset \Diamond \varphi \urcorner$$

or by contraposition

$$\ulcorner \Box \varphi \supset \Box \Box \varphi \urcorner$$

which, when added to the Lewis system S1 (or S2 or S3), or to Gödel's basic system,[27] Feys's system T, or von Wright's system M yields S4 or some equivalent system, e.g., von Wright's system M', and :

$$(5) \qquad \ulcorner \Diamond \varphi \supset \Box \Diamond \varphi \urcorner$$

or by contraposition

---

[27] Postulate sets for all the systems of sentential and modal logic mentioned in this these are given in Appendix I.

$$\ulcorner \Diamond \sim \Diamond \varphi \supset \Diamond \varphi \urcorner$$

which, when added to the Lewis system S1 or any of the other systems mentioned above, yields S5.

In terms of our talk about possible worlds, (4) expresses that if a world X is possible with respect to a world Y, which in turn is possible with respect to a world Z, then X is possible with respect to Z. That is, (4) simply expresses that the relation R is *transitive*:

(4') $(X)(Y)(Z)(XRY.YRZ \mathbin{.\supset.} XRZ)$

Something a little more complicated is expressed by (5), namely that if a world X is possible with respect to world Y, then X is possible with respect to any world Z which is possible with respect to Y. I.e.,

(5') $(X)(Y)(Z)(XRY. ZRY \mathbin{.\supset.} XRZ)$

The property expressed by (5') has no particular name. Granted (3'), (5') does, however, by universal instantiation, imply:

(6') $(Y)(Z)(ZRY \mathbin{.\supset.} YRZ)$

i.e., that the relation R is *symmetrical*.

Granted (3'), (5') also implies (4'). ((4') follows from (5'), using (6') with 'Z' and 'Y' interchanged.) Since on the other hand (4') and (6') together imply (5'), (5') is simply a brief way of stating that the relation R, if totally reflexive, is *symmetrical* and *transitive*. (5') therefore is useful also outside of modal logic, for the characterization of equivalence relations. Instead of saying that a relation R is an equivalence relation if it is reflexive, symmetrical and transitive, i.e., satisfies (3'), (4'), and (6'), we may require it to satisfy just (3') and (5').

Just as (4') and (5') corresponded to (4) and (5) respectively, one should expect that there should be some modal statement corresponding to (6'). Such a statement is:

(6) $\ulcorner \varphi \supset \Box \Diamond \varphi \urcorner$

For, if we start out with world Y, then according to (6) for any world Z which is possible with respect to Y, Y is possible with respect to Z.

Since (4') and (6') are independent, and since, as just noted, they together imply (5'), while (5'), granted (3'), implies (6') and (4'), (6) is useful as an axiom which, when added to S4, yields a system equivalent to S5 without making the special S4 axiom (4) superfluous.

Becker in "Zur Logik der Modalitäten" (1930) proposed to call (6) *"Brouwer's axiom."* In Lewis and Langford's *Symbolic Logic* (1932), the independence of (4) and (6), and the equivalence of the conjunction of (4) and (6) with (5), granted (3), is proved (pp.497-499). But in later comparisons of different systems of modal logic (6) is seldom mentioned, in spite of its virtues from a comparative point of view.

By using "Brouwer's axiom" *instead* of the special S4 axiom one gets a modal system which is incomparable with S4 in strength, but which like S4 is weaker than S5 and stronger than S3 and the Gödel-Feys-von Wright systems. In our discussion of the distinctness of individuals (sections 14 and 15), this new system will play a prominent part.

Chapter Three

Identity of Individuals

12. The Identity of Individuals in Quantified Modal Logic

In this and the following chapters, the difficulties of quantified modal logic will be discussed. As the seemingly queer results we have to accept vary from system to system, it will be helpful to remember that each of the five Lewis systems S1-S5 is stronger than the preceding one, and that Gödel's basic system, Fey's system T, and von Wright's system M are all equivalent, stronger than S2, weaker than S4, incomparable with S3. In the Gödel-Feys-von Wright system and all stronger systems, if φ is provable, so is $\ulcorner \Box \varphi \urcorner$ [1] Let us call this rule 'RL'.

All the systems of quantified modal logic reviewed in section 9 were got from one of these standard systems by adding to it the notation and axioms of quantification theory, and eventually of identity theory or higher order functional calculi. In the only two systems which do not have as a primitive rule

---

[1] This is a primitive rule in the Gödel-Feys-von Wright system. That it is a derived rule in S4 (and S5) was proved by Tarski and McKinsey in "Some theorems about the sentential calculi of Lewis and Heyting" (1948) (Their theorem 2.1, p.5).

of inference that from φ one may infer ⌜□φ⌝, viz., Miss Barcan's and Miss Rasiowa's,[2] an '□' is prefixed to all of these additional axioms.

Only Miss Barcan, Burks, and Fitch added extra axioms to govern the mixture of modal operators and quantifiers. Miss Barcan added,

$$(1) \qquad ⌜\Diamond(E\alpha)\varphi \prec (E\alpha)\Diamond\varphi⌝$$

Burks added,

$$(2) \qquad ⌜(\alpha)\Box\varphi \supset \Box(\alpha)\varphi⌝$$

Fitch discussed the consequences of addition (1) with material, not strict conditional, and also those of adding (2), without remarking that the two are equivalent (by contraposition).

After this digression, let us now examine further our example in section 11 relating to the two objects a and b and two possible worlds. We took '□Fa' to mean that F is true of a in all worlds which are possible with respect to the actual one, and '(Ex)□Fx' to mean that there is some object x of which F is true in all these worlds.

To make sense of this, we need some theory of individuation which accounts for identification of objects from one possible world to another. If, e.g., one wants to keep track of an object from one possible world to another by the help of its attributes, one might, perhaps, restricting oneself for a moment to *uniterated* modalities, hold that every object has some attribute which it keeps in all worlds which are possible with respect to the actual one and which no other object has in any of these possible worlds. Thus, for a in our example, that 'F' is true of it, or perhaps just that it is called 'a' might be such an attribute. Apparently, also the attribute of possibly being K seems to be an attribute of a satisfying this requirement. In general, the condition could then be formulated as follows, if we permit quantification over attributes (using variables 'f', 'g', etc. for this purpose):

$$(3) \qquad (x)(Ef)((y)(\Box(y \text{ has f}) \equiv. x=y) \vee (y)(\Diamond(y \text{ has f}) \equiv. x=y))$$

[2] In Miss Rasiowa's system this rule is, however, a derived rule (Lemma 2.1 of her "Algebraic treatment of the functional calculi of Heyting and Lewis" (p.109)).

If one admits iteration of modal operators, the '□' and the '◇' in (3) may be replaced by any sequence ' ... ◇ ... □ ...' of modal operators:

$$(x)(Ef)(y)(... ◇ ... □ ... (y \text{ has } f) \equiv . x=y)$$

Furthermore, when iteration of modal operators is permitted, a quantifier with modal operators within its scope may in turn be preceded by modal operators. For the interpretation of such a formula we have the following requirement:

(4)   Let F be a formula to be interpreted, and let Q be any quantifier in F which has at least one modal operator within its scope. Let i be the total number of modal operators within whose scope Q is situated. Then in order to interpret F, we must require that.

$$□_{(1)} □_{(2)} ... □_{(m)} (x) (Ef) (y) (... ◇ ... □ ... (y \text{ has } f) \equiv . x=y)$$

where m is the largest number i correlated with any Q in F.

One might perhaps hope that theories of individuation differing from the one sketched above might enable one to quantify into modal contexts also in cases where requirement (4) is not fulfilled. It turns out, however, that a condition even stronger than (4) is fulfilled regardless of one's theory of individuation, viz.,

(5) $$(Ef) (□_{(1)} □_{(2)} ... □_{(m)} (x) (y) (... ◇ ... □ ... (y \text{ has } f) \equiv . x=y))$$
where m is an in (4).

For it may be shown that, regardless of one's theory of individuation, one must have,

(6) $$□_{(1)} □_{(2)} ... □_{(m)} (x) □_{(1)} □_{(2)} ... □_{(n)} (x=x)$$

where m is as above, and, is any natural number (m and n are mutually independent).

Actually, something yet stronger than (6) can be shown to hold, viz.,

(7) $\Box_{(1)} \Box_{(2)} \dots \Box_{(m)}$ (x)(y)(x=y . $\supset \Box_{(1)} \Box_{(2)} \dots \Box_{(n)}$ (x=y))
                    where, as above, m and n are any natural numbers.

As a first step toward proving (7), we will consider the weaker:

(8)   (x)(y)(x=y . $\supset \Box$(x=y))

Suppose that (8) did not hold. In terms of an example like that of section 11, this would mean that we have a situation like:

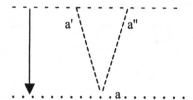

where there correspond to some one object *a* in the actual world two (or more) objects *a'* and *a"* in a world possible with respect to the actual one. If in a situation like this we were to interpret a formula like '(Ex) $\Box$Fx' (or '(x) $\Box$Fx', '(Ex) $\Diamond$Fx', or '(x) $\Diamond$Fx)', we could no longer say about an object in the range of our variables, viz., in our actual world, that *it* has such and such properties, e.g., F, in all possible worlds. Hence we can have no "splitting" of individuals as we pass from the actual world into one which is possible with respect to it. So (8) holds. Neither can any individual "disappear" during this process. In addition to (8) we, therefore, have the following corollary to (8):

(9)   Every object in the actual world exists in every possible world,
            i.e., every object exists necessarily.

If instead of '(Ex)$\Box$Fx', we consider, e.g., '(Ex)$\Box\Box$Fx', parallel reasoning leads us to

(x)(y)(x=y . $\supset\Box\Box$(x=y))

and so on.

And, if we consider, e.g., '$\Box(Ex)\Box Fx$', '$\Box\Box(Ex)\ \Box FX$', etc., we are similarly led to:

$$\Box(x)(y)(x=y \mathbin{.} \supset\Box(x=y))$$
$$\Box\Box(x)(y)(x=y \mathbin{.} \supset\Box(x=y))$$

etc.

Consideration of, for example,

(10) $\Box_{(1)}\ \Box_{(2)} \ldots \Box_{(m)}\ (Ex)\ \Box_{(1)}\ \Box_{(2)} \ldots \Box_{(n)}Fx$

finally gives us (7). So we may conclude:

> *Thesis 1.*   For any interpreted system of quantified modal logic with identity which permits expressions like (10) above, the following is true:
> $\Box_{(1)}\ \Box_{(2)} \ldots \Box_{(m)}\ (x)(y)\ (x=y \mathbin{.} \supset\Box_{(1)}\ \Box_{(2)} \ldots \Box_{(n)}(x=y))$
> where m and n are any two natural numbers.

From Thesis 1 we may infer the corollaries:

> *Corollary 1.1*   For any interpreted system of quantified modal logic (etc.) the following is true:
>
> $\Box_{(1)}\ \Box_{(2)} \ldots \Box_{(m)}\ (x)\Box_{(1)}\ \Box_{(2)} \ldots \Box_{(n)}(x=x)$

> *Corollary 1.2*   For any interpreted system of quantified modal logic (etc.) the following is true:
>
> $\Box_{(1)}\ \Box_{(2)} \ldots \Box_{(m)}\ (x)(Ey)\ \Box_{(1)}\ \Box_{(2)} \ldots \Box_{(n)}(x=y)$

Corollary 1.3   If '$\Box_{(1)}$ $\Box_{(2)}$ $\ldots$ $\Box_{(m)}$ (x)(y) (x=y $.$ $\supset\Box_{(1)}$ $\Box_{(2)}$ $\ldots$ $\Box_{(n)}$(x=y))' (or the formulae in Corollary 1 or 2) is not derivable in a system of quantified modal logic with identity, then the system is (semantically) incomplete.[3]

13. The Identity of Individuals in the Proposed Systems of Quantified Modal Logic

How do the results of the preceding section compare with what one already knows about the existing systems of quantified modal logic?

Already Miss Barcan, in her extension of S2 in "The identity of individuals in a strict functional calculus of second order" (1940) proved the formula:

(1) $(x)(y)(x=y . \supset\Box(x=y))$

with a strict, rather than a material conditional. (Formula (8) of the preceding sections. Theorem 2.31 of Miss Barcan's paper.)

The essential part of her proof may be rephrased as follows in Quine's notation of natural deduction:

*(1')   x=y

---

[3] A formal system S will be said to be semantically complete with respect to an interpretation i if and only if every formula which comes out true under the interpretation i is derivable in S. If every formula which comes out true under every interpretation of S is derivable in S, S will, for short, be said to be *semantically complete*. Very far-off "interpretations" of, e.g., quantifiers, connectives, or modal operators will not qualify as interpretations. We trust, however, that our use of the term 'interpretation', as illustrated in the preceding section, is broad enough to cover every interpretation of quantified modal logic acceptable to modal logicians and philosophers.

For the following discussion, also the notion of *syntactical consistency* will be needed. Since all the systems discussed contain a sign ' $\sim$ ' for negation, the following standard definition will do: S is syntactically consistent if and only if for no formula φ of S both φ and $\ulcorner\neg\varphi\urcorner$ are derivable in S.

*(2')	□(x=x)	
*(3')	□(x=y)	(1') (2') (substitutivity of identity)
(4')	x=y . ⊃□(x=y)	*(3')
(5')	(y)(x=y . ⊃□(x=y))	(4') y
(6')	(x)(y)(x=y . ⊃□(x=y))	(5') x

Line (2') is a theorem of Miss Barcan's (2.6 of her paper). The proof of this is straightforward when one remembers that Miss Barcan has prefixed an '□' to all her axioms of the second order functional calculus (see note 2, section 9).

In the above proof, Miss Barcan does not draw on her special axiom for the mixture, nor on the special axiom ⌜◇(φ.ψ)≺ ◇φ⌝, which distinguishes S2 from S1. We may, therefore, conclude:

(I)   '(x)(y)(x=y . ⊃□(x=y))' is provable in any system got from S1 or any stronger system by adding the notation and axioms of quantification theory and identity theory or of the second order functional calculus (with an '□' prefixed to these axioms if the rule RL: ⊢φ → ⊢ ⌜□φ⌝ is not among the primitive rules of the system).

That either of the two conditions in the parenthesis is sufficient is seem immediately from the structure of the proof just given.[4]

Since S1 is the weakest of the systems of modal logic reviewed at the beginning of the preceding section, '(x)(y)(x=y . ⊃□(x=y))' is provable in the extensions of all of these systems, provided that the extension satisfies the requirements stated in (I).

In fact,

(2) $\square_{(1)} \square_{(2)} \ldots \square_{(m)}$ (x)(y) (x=y . ⊃$\square_{(1)} \square_{(2)} \ldots \square_{(n)}$(x=y))

where *m* and *n* are any natural numbers

(formula (7) of the preceding section) is provable in many of these systems:

---

[4] In (I) and the results which follow when a system is said to be supplemented by the axioms of identity theory, this means, in particular, that the identity relation is substitutive with respect to all contexts expressible in the system. For a discussion of this supposition, see section 18.

(II)   (2) is provable in *any* system of modal logic which in addition
           to S1 or truth functional logic contains the notation and
           axioms of quantification theory and identity theory or of the
           second order functional calculus, provided that the rule RL:
           ($\vdash \varphi \rightarrow \vdash \ulcorner \Box \varphi \urcorner$) is among the primitive or derived rules of the
           system.

The proof is immediate: rule RL is applied n times to the identity axiom (or
derived identity theorem) 'x=x', the result is substituted for line (2') in Miss
Barcan's proof sketched above ((1') – (6')), and rule RL is then applied m times
to line (6') of the new proof.

Since rule RL is a primitive rule in the Gödel-Feys-von Wright systems, this
strong identity result (2) is provable in all (quantifier-identity) extensions of
these, and or all stronger systems, including e.g., von Wright's systems M' and
M".[5] In particular, if one adds quantifiers and identity to the Lewis systems S4
and S5, in which RL is a derived rule, and RL is still derivable in the extension,
then (2) is derivable in the extension.

Of the systems surveyed in section 9, Fitch's intuitionistic modal logic with
quantifiers, Rasiowa's extension of S4 and Burks' logic of causal propositions
do not comprise a theory of identity. (Rasiowa's '=', like that of Lewis, is the
strict biconditional. Fitch, too, has Lewis' strict biconditional in mind when, on
p. 144 of "Intuitionistic modal logic with quantifiers," he says that "the rule
allowing substitution of equals for equals is taken as primitive in S2[1] and S4[1]
[Miss Barcan's system] but is a derived rule in M [Fitch's system].") Therefore,
although the rule RL is a primitive rule in the system of Fitch and Burks, and a
derived rule in that of Miss Rasiowa, (1) and (2) are not discussed by these
authors.

Among the remaining systems, Carnap's two systems, one of the systems
considered in Fitch's *Symbolic Logic*, Bayart's systems, and Kripke's system all
contain RL as a primitive rule.[6]

---

[5] The axioms and rules of M, M', and M" are stated in von Wright's *Essay in
Modal Logic*, Appendix II.

[6] Carnap "Modalities and quantification" T10-1a and T10-1b, Carnap *Meaning
and Necessity* 41-1. Cf., convention 39-3.

Fitch *Symbolic Logic*. Fitch's rule of necessity introduction (11.3) yields RL
as a derived rule in his version of the Lewis calculus S4, but not in his version of

Miss Barcan does not derive RL in any of her two systems. She does, however, prove

$$(3) \qquad \Box(x)(y)(x=y \, . \supset \Box(x=y))$$

in her extension of S4.[7]

In her proof of (3) '$\Box(x=x)$' plays a role similar to the one it plays in line (2') of the proof above. With the help of the special axiom for S4, $\ulcorner\Box\varphi \supset \Box\Box\varphi\urcorner$ and modus ponens one sees then easily that,

$$(2) \qquad \Box_{(1)} \Box_{(2)} \ldots \Box_{(m)} (x)(y) (x=y \, . \supset \Box_{(1)} \Box_{(2)} \ldots \Box_{(n)}(x=y))$$

is provable in Miss Barcan's extension of S4.

Although Miss Barcan does not derive the rule RL in her extension of S4, RL can easily be derived in it.[8] With the help of RL, Miss Barcan's proof of (3)

S2, due to the restriction in the latter on reiteration into subordinate proofs (11, 10).

Bayart's "Correction de la logique modale du premier et du second ordre S5" règle IL, p. 40.

Kripke "A completeness theorem in modal logic" R1. (p.1)

[7] Theorem 2.33*, p. 15 of Miss Barcan's "The identity of individuals in a strict functional calculus of second order."

[8] Rule RL can be derived in Miss Barcan's extension of S4 e.g., along the following lines:

Miss Barcan proves a deduction theorem of the following form in her extension of S4 (using '$\prec$' for the strict conditional, '$\equiv$' for the strict biconditional):

$$(1'') \qquad \text{If} \qquad \varphi_1, \varphi_2, \ldots, \varphi_n \vdash \psi \text{ and if } \vdash \ulcorner\varphi_1 \equiv \Box\chi_1\urcorner, \vdash \ulcorner\varphi_2 \equiv \Box\chi_2\urcorner, \ldots, \vdash \ulcorner\varphi_n \equiv \Box\chi_n\urcorner, \text{ then } \varphi_1, \varphi_2 \ldots \varphi_{n-1} \vdash \ulcorner\varphi_n \prec \psi\urcorner$$

(Theorem XXIX*) of Ruth C. Barcan "The deduction theorem in a functional calculus of first order based on strict implication" (1946), p. 118. As proved by Miss Barcan on p. 117 of her paper, this theorem is not provable in her extension of S2)

Now if a formula φ is provable in Miss Barcan's system, i.e., if

(2") $\vdash \varphi$

then, trivially,

(3") $\ulcorner \Box \psi \vdash \varphi \urcorner$

where $\ulcorner \Box \psi \urcorner$ is any statement. Let $\ulcorner \Box \psi \urcorner$ be provable, i.e.,

(4") $\vdash \ulcorner \Box \psi \urcorner$

By (1") and (3"):

(5") $\vdash \ulcorner \Box \psi \prec \varphi \urcorner$

Since, however, in Miss Barcan's system:

If $\vdash \ulcorner \varphi \prec \psi \urcorner$ then $\vdash \ulcorner \Box \varphi \prec \Box \psi \urcorner$

(her VII), and since by the special axiom of S4:

$\vdash \ulcorner \Box \psi \prec \Box \Box \psi \urcorner,$

we may, by Miss Barcan's axiom 5 (transitivity of strict implication) and her rule I (modus ponens), conclude from (4") and (5"):

$\vdash \ulcorner \Box \psi \urcorner$

The Theorems referred to in this proof of RL are theorems of Miss Barcan's first order functional calculus based on S4. Since, however, as observed by Miss Barcan on p. 13 of "The identity of individuals in a strict functional calculus of second order" (1947), all the proofs of her first order functional calculi based on S2 and S4 can be paralleled in the corresponding extended systems for variables of higher type, we may conclude,

> Rule RL (if $\vdash \varphi$ the $\vdash \ulcorner \Box \varphi \urcorner$) is derivable in Miss Barcan's first and second order functional calculi based on S4.

above can be greatly simplified. (3) would follow directly from (1) by just one application of rule RL.

In Bayart's and Kripke's systems (2) is easily proved along the lines indicated in the proof of Theorem II above.

Fitch, in his *Symbolic Logic*, proves

(1)   a=b . ⊃ □(a=b)

(his theorem 23.4).

Fitch's proof of (4) may easily be strengthened so as to yield (1). In Fitch's symbolism the strengthened proof becomes:

1	x		x=y		hyp
2			□ [x=y]		23.2
3			□ [x=y]		1. 2. id elim
4			[x=y] ⊃□ [x=y]		1-3, imp int
5		(y)	[[x=y] ⊃□[x=y]]		1-4, u q int
6	(x)	(y)	[[x=y] ⊃□[x=y]]		1-5, u q int

Also (2) is provable in Fitch's system along similar lines.[9]

---

[9] The proof of (2) in Fitch's system would run as follows:

1	□ □ ... □	x	y	x=y		hyp
2				□ [x=x]		23.2
3				□	□ [x=x]		2, reit
4				□□ [x=x]		3-3, nec int
5				□	□□ [x=x]		4, reit
2n				□□□ [x=x]		(2n-1)-(2n-1), nec int
2n+1				□□□ [x=y]		1, 2n1 id elim
2n+2				[x=y] ⊃ [□□□ ... □ [x=y]]		1-2n+1, imp int
2n+3			(y)	[[x=y] ⊃ [□□□ ... □ [x=y]]]		1-2n+2, u q int
2n+4		(x)	(y)	[[x=y] ⊃ [□□□ ... □ [x=y]]]		1-2n+3, u q int
2n+5				□ (x) (y) [[x=y] ⊃ [□□□ ... □ [x=y]]]		1-2n+4, nec int
2n+m+3	□ ... □ (x) (y) [[x=y] ⊃ [□□□ ... □ [x=y]]]					1-2n+m+2, nec int
2n+m+4	□ □ ... □ (x) (y) [[x=y] ⊃ [□□□ ... □ [x=y]]]					1-2n+m+3, nec int

In Carnap's system MFC in "Modality and quantification," (1) and (2) are immediate corollaries of his theorem T10-3 f.

In Carnap's system S_2 in *Meaning and Necessity* "we cannot," Carnap says, "simply speak of identity but must distinguish between identity of extension and identity of intension, in other words between equivalence and L-equivalence" (p.100).

Certainly what Carnap calls "identity of extension" does not correspond to what we called "identity" in section 12, where we discussed the identity of the individuals over which we quantify. For, using 's' for Walter Scott, we should then, according to (1) have,

$$(x)(x \equiv s . \supset \Box (x \equiv s))$$

which contradicts theorem 43-1a C of *Meaning and Necessity* (p. 191), according to which,

$$(Ex)(x \equiv s . \sim \Box (x \equiv s))$$

Hence we may conclude:

(III)　　　If in Carnap's system S_2 in *Meaning and Necessity* we were to regard "identity of extension" ('$\equiv$') as corresponding to what we called "identity" in the discussion in section 12, then if S_2 were (semantically) complete, it would by (syntactically) inconsistent.

Apparently, S_2 would be incomplete and consistent. For although S_2 contains the rule RL (*Meaning and Necessity*, 39.3), (1) (and (2)) do not seem to be derivable in it. Carnap's relation '$\equiv$' is substitutive in extensional contexts only.[10] Since $\ulcorner \Box \varphi \urcorner$ is non-extensional with respect to φ (*Meaning and Necessity*

This proof of (2) does not go through if one makes the restriction on reiteration into a strict subordinate proof mentioned in 11.10 of Fitch's Symbolic Logic. As pointed out by Fitch (Symbolic Logic, p. 66) the restricted system would be almost the same as the Lewis system S2.

[10] The two theorems 12-3 and 12-4 concerning interchangeability in *Meaning and Necessity* should be contrasted with D8-1g of "Modalities and

11-6), Carnap would in particular not permit the substitution by which we got (3') from (1') and (2') in the above proof. Miss Barcan's proof of (1) can, therefore, not be reproduced in Carnap's system in *Meaning and Necessity*.

What Carnap calls "identity of intension" has, however, all the features characteristic of what we called "identity" in section 12. In particular, (1) and (2) with '$\equiv$' for '$=$' are derivable in S_2, e.g., along the lines of the proof of theorem (II) above. For '$\equiv$', unlike '$=$', is substitutive with respect to all contexts of S_2 (*Meaning and Necessity*, 12-4 b), and as just mentioned RL is a rule of S_2. '$\equiv$' is hence apparently strong enough to be an identity relation in S_2. One might thing that '$\equiv$' is perhaps too strong, i.e., that for identical objects a and b in S_2's universe of discourse '$a \equiv b$' may be false. But it turns out that no relation weaker than '$\equiv$' can be an identity relation in S_2. For, as pointed out by Quine in one of his letters to Carnap (*Meaning and Necessity*, p.196), since '$(x)(x \equiv x)$' is a theorem of S_2, the members of S_2's universe of discourse are distinct unless they are L-equivalent. For, if a and b belong to S_2's universe of discourse, then from '$(x)(x \equiv x)$' '$a \equiv a$' follows, and, if a and b are identical, then by the substitutivity of identity, '$a \equiv b$' must be true. Therefore,

> (IV)   In Carnap's system S_2 in *Meaning and Necessity,* the relation '$\equiv$', "identity of intension" has all the features of what we called "identity" in section 12. (1) and (2), with '$\equiv$' for '=', are theorems of S_2. Members of S_2's universe of discourse are individuated by the relation '$\equiv$'. They are, therefore, individual concepts, not individuals, properties not classes, and propositions, not truth values.

In this result and the discussion which preceded it we have talked about Carnap's system S_2 just the same way we have talked about the other proposed systems of quantified modal logic. Carnap himself talks about S_2 in a strikingly different way. In particular, he distinguishes value extensions and value intensions of variables (*Meaning and Necessity*, § 10) and uses the expression "universe of discourse" in such a way that it would be erroneous to say about S_2, as we just did, that "the extensions have disappeared from the universe of discourse of the language" (*Meaning and Necessity*, p.199). The recurring point in Carnap's arguments for his position is that "it is not possible for a predicator

---

quantification" (p. 52): "( ) $[(i_k=i_j) \supset (M_k \supset M_j)]$ where M is like M_k except for containing free occurrences of i_j wherever M_k contains free occurrences of i_k."

to possess only an extension and not an intension." (*Meaning and Necessity*, p.199). With this one may agree. But even if one grants that in intensional and extensional contexts alike predicators and other designators cannot posses only an extension and not an intension, one may with Frege doubt that designators have the same extension in intensional contexts as they have in extensional ones. And Carnap does nothing to soothe these doubts. Decisive for our purpose is, however, that Theorems III and IV above, and the discussion leading up to them, are independent of one's attitude to Carnap's method of extension and intension. For nothing in the system S_2 itself compels us to talk about S_2 in a terminology which is different from that which we have used to discuss the other proposed systems of quantified modal logic. As Quine has put it, the duality between the extensional and intensional values of variables on which Carnap's claim rests "is a peculiarity only of a special metalinguistic idiom and not of the object language itself." (*Meaning and Necessity*, 9.196)

In addition to the Lewis systems S1-S5, three other systems of modal logic, S6, S7, and S8 originated with Lewis. S6 and S7 are got by adding the axiom $\ulcorner \Diamond \Diamond \varphi \urcorner$, or 'every statement is possibly possible,' to S2 and S3, respectively. S8 is got by adding the axiom $\ulcorner \Box \Diamond \Diamond \varphi \urcorner$ to S3. S6-S8 are inconsistent with the Gödel-Feys-von Wright system and with S4 and S5.

Obviously,

(V)   (2) is not derivable in any system which is consistent with $\ulcorner \Diamond \Diamond \varphi \urcorner$

and consequently, by Corollary 1.3:

(VI)   Any system of modal logic with quantifiers and identity which is consistent with $\ulcorner \Diamond \Diamond \varphi \urcorner$ is semantically incomplete.

This means in particular that,

(VII)   (2) is not derivable in any consistent extension of S6 (or in any stronger or weaker system)

and by Corollary 1.3:

(VIII)   Any (consistent) extension of S6-S8 with quantifiers and identity, and any system of quantified modal logic with identity which is stronger or weaker than any of these extensions, is semantically incomplete.

One may also prove that,

(IX)　　(2) is not derivable in any extension of S3 or of any stronger system, unless the extension is equivalent to an extension of S4 (or some stronger system).

For Parry has proven that the following is a theorem of S3:

(5)　　　$\ulcorner \Box \Box \varphi \prec . \Box \psi \prec \Box \Box \psi \urcorner$ [11]

and from this and the fact that (2) is an expression of the form $\ulcorner \Box \Box \varphi \urcorner$, (IX) follows.

From (IX) we may conclude by Corollary 1.3 that

(X)　　　Any semantically complete extension of S3 or of any stronger system with quantifiers and identity is equivalent to an extension of S4 (or some stronger system).

This proves nothing about the extensions of S1 and S2, since (5) is provable neither in S1 nor in S2.[12]

One standard way of proving that (2) is not derivable in a system got by adding a set of axioms and rules for quantification theory and identity theory or second order functional calculus to S1 or S2, would be to specify some interpretation for which the axioms of the resulting systems come out true and its rules of inference sound, and nevertheless (2) false. Such an interpretation is, however, not to be hoped for in this case, since (2) according to Thesis 1 comes out true under all interpretations.

---

[11] Parry, "Modalities in the *Survey* system of strict implication" (1939), p.141 (Theorem 32.1).

[12] That (5) is provable neither in S1 nor in S2 is not proved by Parry. It follows, however, from the fact that formulae of type $\ulcorner \Box \Box \varphi \urcorner$ are provable (by rule RL) in the Gödel-Feys-von Wright system, which is stronger than S2, but weaker than S4.

Another way towards this goal which is not blocked by Thesis 1, would be to prove that the corresponding extension of S7 is (syntactically) consistent. By theorem VII we could then even conclude that no such extension of any system weaker than S8 has (2) as a theorem if the extension is consistent.

A simpler proof of this result is, however, forthcoming if we observe that neither of these added axioms and rules for quantification theory and identity theory or second order functional calculus would permit us to infer any formula starting out with '□□ . . .' from a set of formulae, none of which starts out with '□□ . . .'. By inspection one sees that this is the case even if an '□' is prefixed to each of the added axioms, and even if we also add the "Barcan axiom" ⌜◇(Eα)φ ≺ (Eα) ◇φ⌝.

So we may conclude:

(XI)   (2) is not provable in any (syntactically consistent) system got by supplementing S8 or any weaker system with a set of axioms and rules for quantification theory and identity theory or for the second order functional calculus, even if an '□' is prefixed to the added axioms, and even if we also add the "Barcan axiom" ⌜◇(Eα)φ ≺ (Eα) ◇φ⌝.

From (XI) one may conclude not only that Miss Barcan's second order functional calculus based on S2 is semantically incomplete, but in view of (VIII) that,

(XII)   Any syntactically consistent system of quantified modal logic with identity got by supplementing a system S of sentential modal logic with a set of axioms and rules for quantification theory (etc., as in (XI)), is semantically incomplete if
         1. S is stronger than the Lewis system S6 or
         2. S is weaker than the Lewis system S8.

This means in particular that of the systems got by supplementing the eight Lewis systems S1-S8 with quantification theory and identity theory or second order functional calculus, only the ones got from S4 or S5 may be semantically complete.[13]

Theorem XII agrees well with a weaker incompleteness result from sentential modal logic, reached along different lines by Sören Halldén. Halldén proved in 1951 that: "If C is a calculus identical with S1, S3, or any intermediate calculus, or if C is stronger than S3 and weaker than both S4 and S7, then no normal interpretation *i* of C is such that C is complete with respect to *i*."[14]

It also accords with the fact that the only two systems of quantified modal logic with identity which have been proven complete till now, are extensions of S5 (Kripke's and Bayart's, see section 8, note 11). Also Carnap's MFC which is conjectured by Bernays to be complete (Cf. section 8, note 11), is an extension of S5. (Miss Rasiowa's extension of S4 is also a proven complete, but since it is just a system of modal quantification theory, it lacks identity.)

---

[13] Although the above incompleteness proof is based on the formula

$$(2) \qquad \Box_{(1)} \, \Box_{(2)} \ldots \Box_{(m)} \, (x)(y) \, (x{=}y \, . \supset \Box_{(1)} \, \Box_{(2)} \ldots \Box_{(n)}(x{=}y))$$

we might according to Corollary 1.3 just as well have used

$$(2') \qquad \Box_{(1)} \, \Box_{(2)} \ldots \Box_{(m)} \, (x)\Box_{(1)} \, \Box_{(2)} \ldots \Box_{(n)}(x{=}x)$$

or

$$(2'') \qquad \Box_{(1)} \, \Box_{(2)} \ldots \Box_{(m)} \, (x)(Ey)\Box_{(1)} \, \Box_{(2)} \ldots \Box_{(n)}(x{=}y)$$

For the proof of Theorem III, however, (2) is needed, since (2') and (2''), unlike (2) are easily provable even in systems which restrict substitutivity of identity to extensional contexts.

[14] Sören Halldén "On the semantic non-completeness of certain Lewis calculi" (1951), p.129 (theorem 2).

Chapter Four

Distinctness of Individuals

14. The Distinctness of Individuals, and Mixtures of Quantifiers and Modal Operators in Quantified Modal Logic

In section 12 we found that in quantified modal logic, identicals have to be necessarily identical. The reason was that for purposes of interpretation the individuals of the actual world have to keep their identity in all worlds which are possible with respect to the actual world. (Or in general: the individuals over which we quantify have to keep their identity in all worlds which are possible with respect to the world in which we quantify over them.) That is, each individual in the actual world has to have some specific individual corresponding to it in every world which is possible with respect to the actual world. In terms of our illustration from section 12, the following situation is intolerable:

(1)

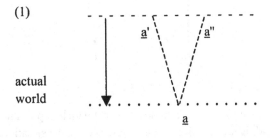

actual
world

53

The following situation appears, however, to be tolerable:

(2)

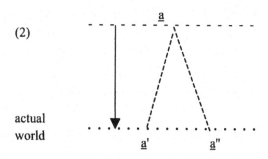

actual
world

i.e., to one object *a* in some world which is possible with respect to the actual one there might correspond two (or more) objects *a'* and *a"* in the actual world. Distinct, or non-identical individuals are hence not necessarily distinct, i.e., we may have

(3) $x \neq y . -\Box(x \neq y)$

So although as argued in section 12 individuals may not "split" as we pass from the actual world in to a world which is possible with respect to it, they may "coalesce" during this process.

Furthermore, whereas individuals may not "disappear" during this passage from one world into one which is possible with respect to it (section 12, result (9)), individuals may "appear" during such a passage. For while the situation:

(4)

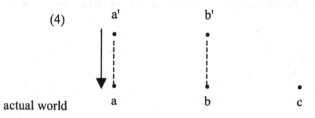

actual world

in which there corresponds no object to *c* in some world which is possible with respect to the actual world, is intolerable, we may well tolerate the situation,

(5)

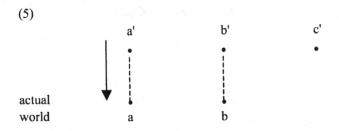

in which some object *c'* in a world possible with respect to the actual one has no object corresponding to it in the actual world.

That is, we may have,

(6) ⌜◇(Eα)φ . –(Eα) ◇φ⌝

Therefore,

(1) ⌜◇(Eα)φ ⊃ (Eα) ◇φ⌝

("Barcan's axiom") and

(2) ⌜(α)□φ ⊃□(α)φ⌝

("Burks's axiom") may be false in some interpreted systems of quantified modal logic.

The converses of (7) and (8), viz.

(9') ⌜(Eα) ◇φ ⊃ ◇(Eα)φ⌝   and
(10') ⌜□(α)φ ⊃ (α)□φ⌝

are, however, bound to come out true in every interpreted system of quantified modal logic.

This follows from the fact that individuals may not disappear as we pass from the actual world to some world which is possible with respect to the actual world.

In fact, since as argued in section 12 individuals may not disappear when we pass from *any* world to one which is possible, or possible possible, etc., with respect to it, (9') and (10') may be strengthened to:

(9) $\ulcorner \Box_{(1)} \Box_{(2)} \ldots \Box_{(m)} ((E\alpha) \Diamond_{(1)} \Diamond_{(2)} \ldots \Diamond_{(n)} \varphi \supset \Diamond_{(1)} \Diamond_{(2)} \ldots$
$\Diamond_{(n)} (E\alpha)\varphi) \urcorner$

and

(10) $\ulcorner \Box_{(1)} \Box_{(2)} \ldots \Box_{(m)} (\Box_{(1)} \Box_{(2)} \ldots \Box_{(n)} (\alpha)\varphi \supset (\alpha) \Box_{(1)} \Box_{(2)} \ldots$
$\Box_{(n)} \varphi) \urcorner$

where m and n are any natural numbers (m and n are mutually independent).
So we have:

Thesis 2.   (9) and (10) are true in every interpreted system of
quantified modal logic.

Corollary 2.1.   Any system of quantified modal logic in which (9) or
(10) is not provable is semantically incomplete

As far as the interpretation of quantifiers in modal logic goes, what we have
to watch out for seems, therefore, to be only what we found already in section 7:

(i)   every individual *i* has to be identified with one and only one
individual in each world which is possible with respect to the
world to which *i* belongs.

With this rule for the *identity* of individuals in mind, let us now find out
what happens when we start putting restrictions on the interrelations between
our possible universes. In section 11 we found that the distinguishing axioms of
S4 and S5, concerning iterated modalities, introduced such restrictions. The
relation R holding between a world *X* and a world *Y* when *X* is possible with
respect to *Y*, was found to be a *transitive* relation when the special axiom of S4,
viz., $\ulcorner \Box\varphi \supset \Box\Box \urcorner$ or $\ulcorner \Diamond\Diamond\varphi \supset \Diamond\varphi \urcorner$, was satisfied. R was found to be
symmetrical when "Brouwer's axiom," viz., $\ulcorner \varphi \supset \Box\Diamond\varphi \urcorner$, holds. And R is
transitive and symmetrical when the special axiom of S5, viz., $\ulcorner -\Box \supset \Box -\Box\varphi \urcorner$ or
$\ulcorner \Diamond \supset \sim\Diamond\sim\Diamond\varphi \urcorner$. Let us see whether any requirements as to the *distinctness* of
individuals are forthcoming when we seek to interpret a system of quantified
modal logic containing any of these formulae.

The special axiom of S4, viz., $\ulcorner \Diamond\Diamond\varphi \supset \Diamond\varphi \urcorner$ or $\ulcorner \Box\varphi \supset \Box\Box\varphi \urcorner$ makes no
change as far as the distinctness of individuals goes. According to this axiom,

the relation R (section 11), is *transitive*. And although R is transitive, individuals may still appear:

(11)

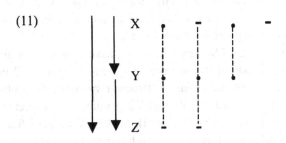

or coalesce:

(12)

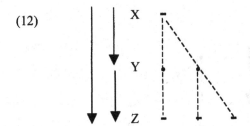

or both:

(13)

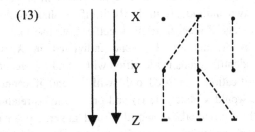

For that any world Y, possible with respect to Z, contains one and only one individual for each individual in Z, and that any world X, possible with respect to Y contains one and only one individual for each individual in Y, is sufficient in order that X contain one and only one individual for each individual in Z. And that X contains only and only one individual for each individual in Z is all that (i) requires in order that X be possible with respect to Z.

"Brouwer's axiom" $\ulcorner \varphi \supset \Box M\varphi \urcorner$ does, however, have consequences for the distinctness of the individuals in the interpreted systems of quantified modal logic which contain this axiom. According to "Brouwer's axiom", the relation R, which holds between two worlds X and Y when X is possible with respect to Y, is *symmetrical* (section 11). According to rule (i), above, as we pass from one world Y to another world X which is possible with respect to Y, each individual in Y must be identified with some one individual in X. Since R is symmetrical, however, also each individual in X must be identified with one individual in Y. One might think that the following situation would satisfy this requirement:

(14)

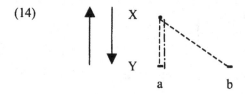

Here, as we pass from world Y to X (dotted lines) the two individuals a and b in Y coalesce into one individual. This certainly does not violate rule (i). Passing from world X to Y (dashed line), the one individual in X becomes identified with one of the two individuals in Y. By itself, neither would this violate rule (i). But the two sets of identifications together give rise to a rather counter-intuitive conception of identity. The one individual in X which according to the first set of identifications is identified with a and b is according to the second set of identification identified only with a, and it cannot be identical with b, since this would violate rule (i) and prevent interpretation of quantifiers ranging over the individuals in world X and governing an open sentence according to which something necessarily or possible is true of the individual(s) in X.

In order to interpret quantifiers in a system containing "Brouwer's axiom" we must, therefore, require that individuals neither "split" *nor "coalesce"* as we pass from one world to one which is possible with respect to it. Neither may individuals "appear" during such a passage from a world X to a world Y, for during the backward passage the individual which appeared in Y would have to be identified with some individual in X. All the individuals in X had, however, by rule (i), already been identified with individuals in Y (and by the very "definition" of 'appearing', with non-appearing individuals in Y).

So we may conclude that,

(15") $\quad$ (x)(y) (x≠y . ⊃□ (x≠y))
(16") $\quad$ (◇(Ex)(Fx) ⊃ (Ex) ◇Fx

and

(17") $\quad$ (x) □Fx ⊃ □(x)Fx

are true in any interpreted system of quantified modal logic with identity in which "Brouwer's axiom" is true.

Taking into account not only worlds which are possible with respect to the one in which we seek to interpret our quantifiers, but also worlds which are possible with respect to these possible worlds, that is, possibly possible, with respect to the world in which we quantify, and so on, (15"), (16"), and (17") may be strengthened respectively to:

(15') $\quad$ (x)(y)(x≠y . ⊃□$_{(1)}$ □$_{(2)}$ ... □$_{(n)}$ (x≠y))
(16') $\quad$ ◇$_{(1)}$ ◇$_{(2)}$ ... ◇$_{(n)}$ (Ex)Fx ⊃ (Ex) ◇$_{(1)}$ ◇$_{(2)}$ ... ◇$_{(n)}$Fx
(17') $\quad$ (x)□$_{(1)}$ □$_{(2)}$ ... □$_{(n)}$ Fx ⊃□$_{(1)}$ □$_{(2)}$ ... □$_{(n)}$(x) Fx

where *n* is any natural number.

In the considerations above, no possible world has been singled out as the actual one. Such a singling out has not been necessary, since the φ in "Brouwer's axiom," as in most other axioms of modal logic, may contain modal operators. There is, therefore, no need to suppose, as in the above formula, that both quantifiers (as in (15')) or one of them (as in (16') and (17')) range over individuals in the actual world. Their range may be any possible world. So we may prefix an arbitrary number of necessity operators to (15'), (16') and (17'), and get:

(15) $\Box_{(1)} \Box_{(2)} \ldots \Box_{(m)} (x)(y)(x=y . \supset \Box_{(1)} \Box_{(2)} \ldots \Box_{(n)} (x=y))$

(16) $\Box_{(1)} \Box_{(2)} \ldots \Box_{(m)} (\Diamond_{(1)} \Diamond_{(2)} \ldots \Diamond_{(n)} (Ex)Fx \supset (Ex) \Diamond_{(1)} \Diamond_{(2)} . \ldots \Diamond_{(n)} Fx$

(17) $\Box_{(1)} \Box_{(2)} \ldots \Box_{(m)} ((x) \Box_{(1)} \Box_{(2)} \ldots \Box_{(n)} Fx \supset \Box_{(1)} \Box_{(2)} \ldots \Box_{(n)} (x)Fx$

where *m* and *n* are any natural numbers.

We may therefore conclude:

Thesis 3.   (16) and (17) are true in any interpreted system of quantified modal logic in which "Brouwer's axiom" $\ulcorner \varphi \supset \Box M\varphi \urcorner$ is true

Corollary 3.1   If "Brouwer's axiom" is provable in a system of quantified modal logic, then, unless (16) and (17) are also provable, the system is semantically incomplete.

Thesis 4.   (15) is true in any interpreted system of quantified modal logic with identity in which "Brouwer's axiom" is true.

Corollary 4.1   If "Brouwer's axiom" is provable in a system of quantified modal logic with identity, then, unless (15) is also provable, the system is semantically incomplete.

Granted that $\ulcorner \varphi \supset M\varphi \urcorner$ holds in the system, the special axiom of S5, $\ulcorner -\Box \varphi \supset \Box -\Box \varphi \urcorner$, implies "Brouwer's axiom" and the special axiom of S4. In fact, as we saw in section 11, the conjunction of the two latter axioms is equivalent to the conjunction of the S5 axiom and $\ulcorner \varphi \supset M\varphi \urcorner$.

The above theses and corollaries, therefore, all apply to S5

15. The Distinctness of Individuals, and Mixtures of Quantifiers and Modal Operators in the Proposed Systems of Quantified Modal Logic

Miss Barcan, in her extension of S2 in "A functional calculus of first order based on strict implication" (1946), proved the formula,

(1) $\ulcorner (E\alpha) \Diamond \varphi \prec \Diamond (E\alpha) \varphi \urcorner$

(her theorem 37). In her simple, three-line proof, no appeal is made to the specific axiom which distinguishes S2 from S1, viz., $\ulcorner \Diamond (\varphi.\psi) \prec \Diamond \varphi \urcorner$, nor to Miss Barcan's special axiom for the mixture, viz., $\ulcorner \Diamond (E\alpha) \varphi \prec (E\alpha) \Diamond \varphi \urcorner$. From (1) Miss Barcan easily derives

(2) $\ulcorner \Box (\alpha) \varphi \prec (\alpha) \Box \varphi \urcorner$

(half of her theorem (39)) again without appealing to the special axiom for S2 or the special axiom for the mixture. Both (1) and (2) are, therefore, derivable in any system got by supplementing S1 or any stronger system by a standard set of axioms for quantification theory (with '$\Box$'s' prefixed if the rule RL is not derivable in the system).

In view of the considerations leading up to theorem (XI) in section 13, we may, however, conclude that the stronger formulae

(3) $\Box_{(1)} \Box_{(2)} \ldots \Box_{(m)} (Ex) \Diamond_{(1)} \Diamond_{(2)} \ldots \Diamond_{(n)} Fx \supset \Diamond_{(1)} \Diamond_{(2)} \ldots \Diamond_{(n)} (Ex)Fx)$

(4) $\Box_{(1)} \Box_{(2)} \ldots \Box_{(m)} (\Box_{(1)} \Box_{(2)} \ldots \Box_{(m)} (x)Fx \supset (x) \Box_{(1)} \Box_{(2)} \ldots \Box_{(n)} Fx)$

the formulae (9) and (10) of the preceding section, mentioned in Thesis 2 and Corollary 2.1, being of the form $\ulcorner \Box \Box \varphi \urcorner$, are not derivable in systems of quantified modal logic based on S8 or weaker systems. So we get the incompleteness result (XII) over again.

(3) and (4) are, however, provable in all systems of quantified modal logic stronger than S2 which have RL as a primitive or derived rule, e.g., Gödel's basic system which has the axioms,

A.1 $\ulcorner \Box \varphi \supset \varphi \urcorner$

A.2 $\ulcorner \Box(\varphi \supset \psi) \supset . \ \Box\varphi \supset \Box\psi \urcorner$

The rule

RL: If $\vdash \varphi$, then $\vdash \Box\varphi$

And the definition,

Def. $\ulcorner \Diamond \varphi \urcorner$ for $\ulcorner \sim\Box\sim\varphi \urcorner$

or the Feys system T, or the von Wright system M, which are equivalent to Gödel's basic system.[1]

Since all the systems of quantified modal logic surveyed in section 9, except Miss Barcan's extension of S2, are stronger than S2 and have RL as a primitive or derived rule, we arrive at the following theorem, anticipated in Thesis 2 and Corollary 2.1:

---

[1] A proof of (4) for m=n=1 might run as follows (in Quine's notation of natural deduction with shortcuts):

*(1')	(x)Fx	
*(2')	Fx	(1')
(3')	(x)Fx $\supset$ Fx	*(2')
(4')	$\Box$((x)Fx$\supset$Fx)	(3') RL
(5')	$\Box$((x)Fx$\supset$Fx) $\supset$. $\Box$(x)Fx$\supset\Box$Fx	A.2
(6')	$\Box$(x)Fx$\supset\Box$Fx	(4') (5')
(7')	(x)($\Box$(x)Fx$\supset\Box$Fx)	(6')
(8')	$\Box$(x)Fx$\supset$(x)$\Box$Fx	(7')
(9')	$\Box$($\Box$(x)Fx$\supset$(x)$\Box$Fx)	(8') RL

A proof of (4) for greater values of m and n is obtained from this proof by repeated applications of RL, A.2, and modus ponens after line (6'), and applications of RL to the result.

(3) may be proven along similar lines, or more directly from (4).

(XIII)   (3) and (4) ((9) and (10) of the preceding section) are derivable in all the systems of modal logic surveyed in section 9, except Miss Barcan's extension of S2.

The converses of (3) and (4), viz.,

(5) $\Box_{(1)} \Box_{(2)} \ldots \Box_{(m)} (\Diamond_{(1)} \Diamond_{(2)} \ldots \Diamond_{(n)} (\text{Ex}) \text{Fx} \supset (\text{Ex}) \Diamond_{(1)} \Diamond_{(2)} \ldots \Diamond_{(n)} \text{Fx})$

and

(6) $\Box_{(1)} \Box_{(2)} \ldots \Box_{(m)} ((x) \Box_{(1)} \Box_{(2)} \ldots \Box_{(n)} \text{Fx} \supset \Box_{(1)} \Box_{(2)} \ldots \Box_{(n)} (x) \text{Fx})$

((16) and (17) of the preceding section) are provable in Miss Barcan's extension of S2 for m=n=1, since (5), with m=n=1, is one of her axioms, and (6) is easily derivable from (5) (Miss Barcan's extension of S4 (Cf., Section 13, note 8), and (5) and (6) for arbitrary *m* and *n*, are derivable in this extension. For by repeated application of RL and (2) above to Miss Barcan's theorem 39 and thereafter of RL to the result, we get (6), from which (5) easily follows.

Since also Burks' system contains RL and the special axiom $\ulcorner(\alpha)\Box\varphi \supset \Box(\alpha)\varphi\urcorner$ for the mixture, (5) and (6) are derivable in his system.

Also in Carnap's system MFC, his extension of S5 in "Modalities and quantification" (1946) (with m=n=1) is adapted as an axiom.[2] In his system S_2 in *Meaning and Necessity*, however, Carnap states (5) and (6) (with m=n=1) as theorems.[3]

The formula,

(7) $\Box_{(1)} \Box_{(2)} \ldots \Box_{(m)} (x)(y)(x \neq y . \supset \Box_{(1)} \Box_{(2)} \ldots \Box_{(n)} (x \neq y))$

---

[2] Axiom D10-1, 1 on p.54 of "Modalities and quantification" (1946). Also (2) is adopted as an axiom by Carnap (D10-1, m).

[3] *Meaning and Necessity* (1947), p. 186, theorems 41-5 k and 41-5 m, respectively.

(15) of the previous section, is derived in only one of the systems surveyed in section 9, viz., in Carnap's system MFC.[4]

In 1955, Prior in his *Formal Logic* (p. 206) proved (7) in S5, supplemented by a standard system of quantifiers and identity.

The following year, in "Modality and quantification in S5" Prior presented a proof of (5) in S5 supplemented by quantification theory. As an appendix to this proof, Prior also proved that the rule of inference

$$(8) \qquad \text{If } \vdash \ulcorner \varphi \supset \Box \psi \urcorner \text{ then } \vdash \ulcorner \Diamond \varphi \supset \psi \urcorner$$

holds in S5.[5] In his review of Prior's article Alan Ross Anderson noticed that also the converse holds in S5, viz.:

$$(9) \qquad \text{If } \vdash \ulcorner \Diamond \varphi \supset \psi \urcorner \text{ then } \vdash \ulcorner \varphi \supset \Box \psi \urcorner\, [6]$$

In what follows, it will be proven that all the theorems (5) (and (6)), (7), (8), and (9) hold in a quantified modal logic with identity based on a system weaker than S5,[7] viz., the system got by adding to the Gödel system above "Brouwer's axiom"

$$\text{A.3} \qquad \ulcorner \varphi \supset \Box \Diamond \varphi \urcorner$$

The system is incomparable in strength with S4.

---

[4] It is an immediate consequence of his theorem T10-3 f, on p. 57 of "Modalities and quantification" (1946).

[5] A.N. Prior, "Modality and quantification in S5" (1956), p. 62.

[6] Alan Ross Anderson, "Review of Prior's 'Modality and quantification in S5'" *Journal of Symbolic Logic,* 22 (1957), p. 91

[7] One sees easily that A.3 is a theorem of S5:

$$
\begin{array}{lll}
(1) & \vdash \ulcorner \Diamond \varphi \supset \Box \Diamond \urcorner & \text{Special axiom of S5} \\
(2) & \vdash \ulcorner \varphi \supset \Diamond \urcorner & \text{A.1} \\
(3) & \vdash \ulcorner \varphi \supset \Box \Diamond \urcorner & (1)\,(2)
\end{array}
$$

That the system is actually weaker than S5 and incomparable in strength with S4 is proved by Lewis in Lewis and Langford, *Symbolic Logic*, p. 498.

Let us first prove (9):

(1')	$\Diamond p \supset q$	Assumed to be provable
(2')	$\Box(\Diamond p \supset q)$	(1') RL
(3')	$\Box(\Diamond p \supset q) \supset . \ \Box\Diamond p \supset \Box q$	A.2
(4')	$p \supset \Box\Diamond p$	A.3
*(5')	p	
*(6')	$\Box q$	(2')(3')(4')(5')
(7')	$p \supset \Box q$	*(6')

(8) then follows easily:

(1')	$p \supset \Box q$	Assumed to be provable
(2')	$\sim\Box q \supset \sim p$	(1')
(3')	$\Diamond\sim q \supset \sim p$	(2') Def
(4')	$\sim q \supset \Box\sim p$	(3') by rule (9), which was just derived
(5')	$\sim\Box\sim p \supset q$	(4')
(6')	$\Diamond p \supset q$	(5') Def

With the help of rule (9), (7) may be proven as follows:

(1')	$x=y . \supset \Box(x=y)$	proved in section 13 (theorem II)
(2')	$\sim\Box(x=y) \supset \sim(x=y)$	(1')
(3')	$\Diamond(x \neq y) \supset . \ x \neq y$	(2') Def
(4')	$x \neq y . \supset \Box(x \neq y)$	(3') by rule (9)
(5')	$(y)(x \neq y . \supset \Box(x \neq y))$	(4') y
(6')	$(x)(y)(x \neq y . \supset \Box(x \neq y))$	(5') x

(7), with arbitrary *m* and *n*, follows easily by starting out with '$x=y . \supset \Box_{(1)} \Box_{(2)} \ldots \Box_{(n)} (x=y)$' and repeated use of rule (9). RL is then applied *m* times to the last line of the resulting proof.

Also Prior's main result in S5, viz. formula (5), concerning quantifiers and modal operators, comes easily in this weaker system with the help of rules (8) and (9):

*(1')	$\Diamond Fx$	
*(2')	$(Ex)\,\Diamond Fx$	(1)
(3')	$\Diamond Fx \supset (Ex)\,\Diamond Fx$	*(2')
(4')	$Fx \supset \Box(Ex)\,\Diamond Fx$	(3') by rule (9)
(5')	$(x)(Fx \supset \Box(Ex)\,\Diamond Fx)$	(4') x
(6')	$(Ex)Fx \supset \Box(Ex)\,\Diamond Fx$	(5')
(7')	$\Diamond\,(Ex)Fx \supset (Ex)\,\Diamond Fx$	(6') by rule (8)

Theorem (5) with arbitrary m and n may be got from the above proof by starting out with '$\Diamond_{(1)}\,\Diamond_{(2)} \ldots \Diamond_{(n)}\,Fx$' and using the rules (9) and (8) repeatedly. RL is then applied m times to the last line of the resulting proof.

Theorem (6) may be proven along similar lines, or directly from theorem (5).

We have hereby proved the results anticipated in Thesis 3, Corollary 3.1, Thesis 4, and Corollary 4.1:

(XIV)   (5) and (6) ((16) and (17) of the preceding section) are provable in a system got by supplementing a standard system of quantification theory by Gödel's basic system + "Brouwer's axiom."

(XV)   (7) ((15) of the preceding section) is provable in a system got by supplementing a standard system of quantification theory and identity by Gödel's basic system + "Brouwer's axiom."

Since "Brouwer's axiom" is a theorem in S5, the above results hold *a fortiori* in systems based on S5. They hold for all extensions of S5 surveyed in section 9, but are stated as theorems by Carnap only. If, however, in Carnap's system S_2 in *Meaning and Necessity* we were to regard '$\equiv$', rather than '$\cong$', as an identity sign, (7) would no longer be provable. For Carnap's theory of identity in S_2 would then as we observed in section 13, not be a standard one, and line (1') in the proof of (7) above would not be provable in S_2 (Section 13, Theorem III).

Chapter Five

Singular Terms and Singular Inference

16. Definite Descriptions in Modal Logic

In section 12 we found that the formula

(1) $(x)(y)(x=y . \supset \Box(x=y))$

is derivable in any semantically complete system of quantified modal logic with identity. In section 13 this result was used to establish that Carnap's system of modal logic in *Meaning and Necessity* is semantically incomplete if '$\equiv$' is construed as an identity sign (Theorem III). We mentioned there that if S is a semantically complete system of quantified modal logic with identity, in which descriptions are treated the Fregean, not the Russellian way, as singular terms which in particular have access to identity contexts and may be used to instantiate universal quantifications, S contains among its theorems:

(2) $\ulcorner (\imath\alpha)\varphi = (\imath\alpha)\,\psi . \supset \Box[(\imath\alpha)\,\varphi = (\imath\alpha)\,\psi]\urcorner$

No (2) is an undesirable result. Let us consider our example from section 11 again, with two possible worlds, and two objects, *a* and *b*:

possible,		
non-actual	Fa. ~Ga. ~ Ha. Ka	~Fb.Gb.Hb. ~Kb
world		
actual world	Fa. ~Ga.Ha. ~Ka	~Fb.Gb.~Hb.~Kb

This example satisfies the standard axioms and rules of inference of quantification theory and identity theory, supplemented by e.g., S5 or any weaker system of modal logic. But the example does not permit a theory of descriptions treating descriptions as names, since in the example we would then have e.g.,

$$(3) \qquad (\imath x)Hx = (\imath x) \ Fx \ . \ \text{-}\Box[(\imath x) \ Hx = (\imath x) \ Fx]$$

which contradicts (2).

Not all philosophers, however, regard (2) as undesirable. Thus, discussing descriptions in modal logic, Feys writes in "Les systèmes formalizes des modalités aristotéliciennes" (1950):

> Mais, n'y a-t-il pas deux sortes d'identité, i et I? Et n'y a-t-il donc pas deux sortes de descriptions, de sorte que ce qui fait un objet unique dans un sens ne fait pas un objet unique dans l'autre? Pour reprendre un exemple autour duquel a été écrite une imposante série d'articles, "l'Étoile du Soir" et "l'Étoile du Matin" se trouve "matériellement" être le même aster, Vénus; elles ont donc, en fait, les mêmes attributs. Mais ces mêmes attributs n'appartiennent pas a l'Étoile du Soir et a l'Étoile du Matin en vertu d'une nécessité logique. Donc, du point de vue de l'identité i, les deux étoiles no font qu'un; du point de vue de l'identité I elles sont deux. De même Monsieur *x* apparaissant en qualité de *a* et Monsieur *x* apparaissant en qualité de *b* pourront devoir être identifies du point de vue de l'identité i et distingués du point de vue de l'identité I.
>
> Cette situation paradoxale existe en effet pour les systèmes plus faibles que S4[1], et, pour cette raison comme pour d'autres, ces

---

[1] This slip of Feys's is apparently due to too hasty reading of Miss Barcan's "The identity of individuals in a strict functional calculus of second order"

systèmes mériteraient d'être étudiés plus amplement. Mais dans les systèmes sur lesquels l'attention s'est jusqu'ici principalement portée, S4 et S5, la question ne se pose pas, puisque i et I s'identifient. (pp. 507-8)

But there are other, harder-to-digest consequences of treating descriptions as proper names in modal logic. Let S be an interpreted system of modal logic which contains the axioms and rules of Gödel's basic system, supplemented by a standard set of axioms and rules for quantification theory and identity theory, and a standard theory of descriptions[2], in which descriptions are defined the Carnap defines them in *Meaning and Necessity*:

(4)   '- -$(\imath x)$ (. . x . .)- -'

for

'(Ey) (x)[(. . x . . $\equiv$. x=y) . - - y - -] v.
~(Ey) (x) (. . x . . $\equiv$. x=y) . - - a* - -'

(*Meaning and Necessity*, 8-2. a* is some arbitrary individual, selected as a common descriptum for all descriptions in which ' . . x . . ' is true of no object or of more than one object.)

Then one may prove that if among the objects of our universe there is one, z, which is necessarily distinct from a*, then any true sentence of the system S is necessarily (necessarily necessarily, etc.) true.

In the proof we will find it convenient to make use of the theorem:

(5) $\ulcorner \Box \varphi . \ \Box \psi . \supset . \ \Box(\varphi.\psi) \urcorner$

(1947) to which he refers. As we saw in section 13, Miss Barcan proved the two kinds of identity to be equivalent already in S2, and her proof holds also in S1.

[2] As mentioned already in note 4 in Chapter Three, a theory of identity will be considered standard only if, in particular, the identity relation is *universally* substitutive. A theory of descriptions will be said to be standard only if, in particular, any well-formed sentence with 'x' as the only free variable is permitted to take the place of 'Fx' in '$(\imath x)Fx$'.

which, as proved by Lewis,[3] holds already in S1 and *a fortiori* in Gödel's basic system.

The proof that modal distinctions collapse in S under the above conditions runs as follows:

*(1')	(ɿx) (x=z.p) = z.z≠a*	
*(2')	(Ey) [(x) (x=z.p. ≡.x=y).y=z]v.	(1') by def. (4) above
	~(Ey) (x) (x=z.p. ≡.x=y).a*=z	
*(3')	Ey) [(x) (x=z.p. ≡.x=y).y=z]	(1'), (2')
*(4')	(x) (x=z.p. ≡.x=y).y=z	(3') y
*(5')	y=z.p.=.y=y	(4')
*(6')	p	(5')
(7')	(ɿx) (x=z.p) = z.z≠a*.⊃p	*(6')
(8')	□ ((ɿx) (x=z.p) = z.z≠a*.⊃p)	(7') RL
*(9')	p.□(z≠a*)	
*(10')	x=z.p. ≡.x=z	(9')
*(11')	(x) (x=z.p. ≡.x=z)	(10') x
*(12')	(x) (x=z.p. ≡.x=z) .z=z	(11')
*(13')	(Ey) ((x) (x=z.p. ≡.x=z) .y=z)	(12')
*(14')	(Ey) ((x) (x=z.p. ≡.x=y) .y=z)v.	
	~ (Ey) (x) (x=z.p. ≡.x=y).a*=z	(13')
*(15')	(ɿx)(x=z.p)=z	(14') by def. (4) above
*(16')	(x) (y) (x=y. ⊃ □ (x=y))	Theorem I or II
*(17')	(y) ((ɿx) (x=z.p)=y. ⊃□((ɿx) (x=z.p)=y))	
		(16')
*(18')	(ɿx) (x=z.p)=z. ⊃□((ɿx) (x=z.p)=z)	(17')
*(19')	□((ɿx) (x=z.p)=z). □(z≠a*).⊃	
	□((ɿx) (x=z.p)=z. z≠a*)	Theorem (5) above
*(20')	□((ɿx) (x=z.p)=z. z≠a*.⊃p)⊃.	
	□((ɿx) (x=z.p)=z. z≠a*)⊃□p	A.2
*(21')	□p	(8') (9') (15') (18') (19') (20')
(22')	p. □(z≠a*)⊃□p	*(21')

---

[3] Lewis and Langford, *Symbolic Logic*, theorem 19.81.

So if we pick for our z any object necessarily distinct from a*, then any true sentence turns out to be necessarily true. In the above proof, 'p' may represent any sentence. Thus we may conclude not only,

$$p. \ \Box(z{\neq}a^*){\supset}\Box p$$

but also

$$\Box p. \ \Box \ (z{\neq}a^*){\supset}\Box\Box \ p$$
$$\Box\Box p. \ \Box \ (z{\neq}a^*){\supset}\Box\Box\Box p$$
etc.

By truth-functional logic we then have,

$$(6) \qquad p.\Box(z{\neq}a^*){\supset}\Box_{(1)} \ldots \Box_{(m)}p$$

for any natural number m.[4]

In a Russellian theory of descriptions, the condition '$\Box \ (x{\neq}a^*)$' (or '$z{\neq}a^*$') could be dropped altogether, and we might conclude simply

$$(6'') \qquad p{\supset}\Box_{(1)} \ldots \Box_{(m)}p$$

for any natural number m.

So we may conclude:

(XVI) Let S be an interpreted system of modal logic got by supplementing Gödel's basic system (or a stronger system, e.g., S4 or S5) by a standard set of axioms and rules for quantification theory with identity. Then if S contains a standard theory of descriptions (i.e., a theory of descriptions in which any well-formed sentence of the system with 'x' as the only free variable is permitted to take the place of 'Fx' in '(ɪx)Fx'), then

1) If descriptions are treated as names along the lines of Carnap's *Meaning and Necessity*, and if '$\Box \ (z{\neq}a^*)$' is true for some z, or

---

[4] This observation, which simplifies and strengthens my original result, is due to Professor Quine.

if "Brouwer's axiom" $\ulcorner \varphi \supset \square \Diamond \varphi \urcorner$ is a theorem of S and 'z≠a*' is true, then modal distinctions collapse in S.

2) If a Russellian theory of descriptions is used, and if, as is usual in extensional logic, $\ulcorner (\imath \alpha)\varphi \urcorner$ is treated as a name when $\ulcorner (E\beta)$ $(\alpha) (\varphi \equiv .\beta = \alpha)\urcorner$ is true, then modal distinctions collapse in S.

Similar awkward consequences follow, of course, in modal systems containing a theory of classes in which class abstracts are defined with the help of descriptions of the type above. Disastrous consequences follow also if one treats class abstracts as names, adopts the following identity condition for classes"

$$(7) \qquad \alpha = \beta . \equiv . (x)(x\varepsilon\alpha. = .x\varepsilon\beta)$$

and allows substitutivity of concretion, viz., that,

$$(8) \qquad \text{'} x\varepsilon\hat{z} (\ldots z \ldots)\text{'} \text{ and '} \ldots x \ldots \text{'}$$

may replace each other in all contexts.

When Gödel's basic system or a stronger system of sentential modal logic is supplemented by such a system of class theory, modal distinctions collapse unconditionally, as is seen from the following proof:

*(1')	$\hat{x} (x=x.p) = \hat{x} (x=x)$	
*(2')	$(x)(x\varepsilon\hat{x}(x=x.p). \equiv .x\varepsilon\hat{x}(x=x))$	(1') by (7)
*(3')	$x\varepsilon\hat{x}(x=x.p). \equiv .x\varepsilon\hat{x}(x=x)$	(2')
*(4')	$x=x.p. \equiv .x=x$	(3') by (8)
*(5')	p	(4')
(6')	$\hat{x} (x=x.p) = \hat{x}(x=x). \supset p$	*(5')
(7')	$\square(\hat{x} (x=x.p) = \hat{x} (x=x). \supset p)$	(6') RL
*(8')	p	
*(9')	$x=x.p. \equiv .x=x$	(8')
*(10')	$x\varepsilon\hat{x} (x=x.p). \equiv .x\varepsilon\hat{x} (x=x)$	(9') by (8)
*(11')	$(x) (x\varepsilon\hat{x}(x=x.p). \equiv .x\varepsilon\hat{x} (x=x))$	(10') x
*(12')	$\hat{x} (x=x.p). \equiv .\hat{x} (x=x)$	(11') by (7)
*(13')	$(x) (y) (x=y. \supset \square (x=y))$	Theorem I (or II)

*(14')   (y) ($\hat{x}$ (x=x.p)=y. $\supset$ $\Box$($\hat{x}$ (x=x.p)=y))

                                                                    (13')

*(15') $\hat{x}$ (x=x.p)= $\hat{x}$ (x=x). $\supset$
           $\Box$($\hat{x}$ (x=x.p)= $\hat{x}$ (x=x))   (14')

*(16') $\Box$($\hat{x}$ (x=x.p)=$\hat{x}$ (x=x). $\supset$p) $\supset$.
           $\Box$($\hat{x}$ (x=x.p)=$\hat{x}$ (x=x)) $\supset$$\Box$p   Gödel's axiom A.2

*(17') $\Box$p   (7') (12') (15') (16')

(18')   p$\supset$$\Box$p   *(17')

By reasoning parallel to that which let to (6) above, we arrive at,

(9)   p$\supset$$\Box_{(1)}$ . . . $\Box_{(m)}$p

We may therefore conclude:

(XVII) Let S be a system of modal logic got by supplementing
Gödel's basic system (or a stronger system e.g., S4 or S5) by a
system of class theory which contains a standard theory of
identity (so that, in particular, the identity relation is
substitutive with respect to all contexts expressible in S), and
such that classes are identical if they have the same members.
Then if in S class abstracts are treated as names and
substitutivity of concretion is allowed, modal distinctions
collapse in S.

Among the proposed systems of modal logic, the second order functional
calculi of Miss Barcan and Bayart permit class abstraction, but do not treat class
abstracts as names. (Their class abstracts cannot be used to instantiate universal
generalizations, and do not have access to identity contexts.)

Carnap's system S_2 in *Meaning and Necessity* is the only proposed system
of modal logic which incorporates a theory of descriptions. In S_2 descriptions
may be used to instantiate universal generalizations and have access to '≡'
contexts and to '≡' contexts.

The disastrous result of XVI above is, however, avoided in S_2 regardless of
whether we construe the identity relation as '≡' or '≡'.

If we were to consider '≡' as an identity sign, then as we saw in section 13,
S_2 is incomplete since,

(1) $(x)(y)(x \equiv y. \supset \Box(x \equiv y))$

is not among its theorems. But (1) is presupposed in the proof of theorem XVI
(line (16')). (1) would have been derivable in S_2 if '$\equiv$' had been *universally*
substitutive. So S_2 is saved from a collapse of modal distinctions simply because
identity of extension lacks this most characteristic feature of an identity relation.

What Carnap calls "identity of intension" behaves, however, as we saw in
section 13, exactly the way an identity relation is supposed to do. Nevertheless
the disastrous result in XVI is avoided in S_2 when we use ' $\equiv$.' for the '='. For
in the proof of XVI, the description '$(\imath x)(x \equiv z.p)$' occurs in several of the lines.
And since '$x \equiv z$' is short for '$\Box(x \equiv z)$' (*Meaning and Necessity*, 39-6),
descriptions of this type are not permitted by Carnap, who on p.184 of *Meaning
and Necessity* states that, "in order to avoid certain complications, which cannot
be explained here, it seems advisable to admit in S_2 only descriptions which do
not contain '$\Box$'." Since in any proof of theorem XVI we must be able to prove
that 'p' is true if and only if the entity described in a description is necessarily
distinct from a*, it seems inevitable that at least one description in the proof
contains the sign ' $\equiv$ ', i.e., an '$\Box$'.

So although S_2 contains a theory of descriptions, S_2 is saved from a collapse
of modal distinctions by the circumstance that its theory of descriptions is not a
standard one. (Cf. footnote 2 of this chapter.)

17. Singular Terms and the Individuation of Our Objects

In section 13 we observed that in Carnap's system S_2 the variables take as
their values individual concepts, not individuals, properties, not classes, and
propositions, not truth values (Theorem IV).

One might be tempted to think that an intensional ontology is a price we
have to pay for quantifying into modal contexts. But this is not so. The variables
and quantifiers by themselves do not require us to individuate our entities finer
than they are individuated in extensional logic. The trouble comes in with our
definite singular terms. If, e.g., the descriptions 'the Morning Star' and 'the
Evening Star' are among our singular terms, and if in some world which is

possible with respect to the actual world these two descriptions refer to distinct entities, then by,

(1) $(x)(y)(x=y. \supset \Box(x=y))$

they have to refer to distinct entities also in our actual world. And the material object to which they both refer in non-modal logic therefore has to be expurgated from our universe of discourse. In order to avoid difficulties of this type, the only kind of descriptions which we may treat like singular terms are those descriptions $\ulcorner (\iota\alpha)\varphi \urcorner$ in which φ satisfies the condition:

(2) $\ulcorner (E\beta) \Box(\alpha) (\varphi \equiv. \alpha = \beta) \urcorner$

or in general, if we admit iterated modalities:

(2'). $\ulcorner \Box_{(1)} \ldots \Box_{(m)} (E\beta) \Box_{(1)} \ldots \Box_{(n)} (\alpha)(\varphi \equiv. \alpha = \beta) \urcorner$

where the variables range over only the objects we want to admit in our universe of discourse, i.e., *extensions*. Abstracts, whether defined by descriptions or directly, are subject to parallel restrictions.

Evidently words like 'Hesperus' and 'Phosphorus' give rise to similar difficulties. A natural way out of these difficulties would be to treat such name-like words just the way we would treat descriptions. We replace them in context using contextual definitions of the Russellian type, insisting with Quine on the "primacy of predicates" (*Methods of Logic*, p.218). Instead of saying that Hesperus and Phosphorus are identical we might, e.g., say that there is one and only one object which Hesperizes and one and only one object which Phosphorizes, and these objects are identical.

This solution leads us to regard a word as a proper name of an object only if it refers to this one and the same object in all possible worlds. This does not seem unnatural. Neither does it seem preposterous to assume as we just did, that if a name-like word does not stick to one and the same object in all possible worlds, the word contains some descriptive element.[5] Frege held that "every grammatically well-formed expression representing a proper name always has a

---

[5] This attitude towards proper names is not unlike that of Wilson e.g., in *The Concept of Language*.

sense."[6] And Quine, by insisting on the primacy of predicates, [7] eliminates *all* definite singular terms along the above lines. Russell, in "Knowledge by acquaintance" held that 'I' and 'this' are the only two words "which do not assign a property to an object, but merely and solely name it."[8] In a footnote added in 1917 Russell excluded 'I' from this list. The only remaining word 'this' will hardly be used as a definite singular term in any system of logic, modal or non-modal.

The name-like words which disturbed Russell were the ones which fail to refer to one and only one object in our actual world. The ones which disturb us in quantified modal logic are the ones which fail to refer to one and the same object in all possible worlds. And the solution outlined above strongly resembles Russell's. Russell permitted a description $\ulcorner(\iota\alpha)\varphi\urcorner$ to be treated as a singular term if and only if the following condition is satisfied:

$$(3) \qquad \ulcorner(E\beta)(\alpha)\ (\varphi\equiv.\ \alpha=\beta)\urcorner$$

In modal logic, we had to require:

$$(2) \qquad \ulcorner(E\beta)\ \Box(\alpha)\ (\varphi\equiv.\ \alpha=\beta)\urcorner$$

or rather the stronger (2'). If and only if this condition is satisfied can $\ulcorner(\iota\alpha)\varphi\urcorner$ be treated as a singular term.

If we want to keep extensional objects in our universe of discourse in a system of quantified modal logic three main approaches are now open to us:

1)   We eliminate all definite singular terms e.g., by insisting on the primacy of predicates and replacing them in context as illustrated above in connection with Hesperus and Phosphorus example. ('Fa' is replaced by '(Ey) [(x)(x a's≡.x=y).Fy]', etc. How much of the context we should include in 'F', i.e., the scope of the "description", would be a problem on a par with other problems raised by logical paraphrase.)

---

[6] "Über Sinn und Bedeutung," quoted from Black's translation in *Philosophical Writings of Gottlob Frege*, p.58.

[7] *Methods of Logic*, p.218.

[8] "Knowledge by acquaintance", in *Mysticism and Logic*, p. 224.

2)   We permit some singular terms to appear in well-formed expressions of our modal system, viz., those expressions μ which satisfy the condition $\ulcorner(E\beta)\ \Box(\alpha)(\alpha\mu's\equiv.\ \alpha=\beta)\urcorner$ or, if we permit iterated modalities, the stronger condition $\ulcorner\Box_{(1)}\dots\Box_{(m)}$ $(E\beta)\Box_{(1)}\dots\Box_{(n)}\ (\alpha)(\ \alpha\mu's\equiv.\alpha=\beta)\urcorner$. Among these singular terms there may well be descriptions and class abstracts. For descriptions the condition just stated is equivalent to condition (2') above.

3)   We use a full theory of descriptions. Descriptions are eliminated by standard Russellian contextual definitions. Thus, e.g., $\ulcorner(\imath\alpha)\varphi=\beta\urcorner$ expands into $\ulcorner(\alpha)\ (\varphi\equiv.\ \alpha=\beta)\urcorner$ etc. We do, however, not treat a description $\ulcorner(\imath\alpha)\varphi\urcorner$ as a name unless the supporting lemma,

$$\ulcorner(E\beta)\ \Box(\alpha)\ (\varphi\equiv.\ \alpha=\beta)\urcorner$$

is at hand.

Thus one of our laws for substitution of a description for a singular term becomes:

If ψ' is like ψ except for containing free occurrences of $\ulcorner(\imath\alpha)\varphi\urcorner$ in place of some free occurrences of ζ, then,

$\vdash \ulcorner(E\beta)\ \Box(\alpha)\ (\varphi\equiv.\ \alpha=\beta)\supset:(\imath\alpha)\varphi=\zeta\ .\supset.\psi=\psi'\urcorner$

A corresponding law for substitution of a description for a description might be:

If ψ' is like ψ except for containing free occurrences of $\ulcorner(\imath\alpha_1)\varphi_1\urcorner$ in place of some free occurrences of $\ulcorner(\imath\alpha_2)\varphi_2\urcorner$, then,

$\vdash \ulcorner(E\beta_1)\ \Box(\alpha_1)\ (\varphi_1\equiv.\ \alpha_1=\beta_1).$
$(E\beta_2)\ \Box(\alpha_2)\ (\varphi_2\equiv.\ \alpha_2=\beta_2).\supset:$
$(\imath\alpha_1)\varphi_1=(\imath\alpha_2)\varphi_2\ .\supset.\ \psi\equiv\psi'\urcorner$

The antecedent of the main conditional here is, however, unnecessarily strong. It obviously suffices that in each possible world either both descriptions lack a descriptum or they have a common descriptum. This will happen if φ_1 and φ_2 are analytically equivalent. The antecedent of the main conditional therefore may be replaced by $\ulcorner \Box(\varphi_1 \equiv \varphi_2) \urcorner$. So if ψ and ψ' are as just described we have,

$$\vdash \ulcorner \Box(\ \varphi_1 \equiv \varphi_2) \supset: (\imath\alpha_1)\varphi_1 = (\imath\alpha_2)\varphi_2 \ . \supset .\psi = \psi' \urcorner$$

It may therefore happen that two descriptions can be substituted for each other although neither of the two behaves like a name.

For universal instantiation we get:

If χ is like ψ except from containing free occurrences of $\ulcorner(\imath\alpha)\varphi\urcorner$ wherever ψ contains free occurrences of γ, then,

$$\ulcorner (E\beta) \ \Box(\alpha) \ (\varphi \equiv. \ \alpha = \beta) \supset (\gamma) \ \psi \supset \chi \urcorner$$

Class abstracts are treated along similar lines. All name-like expressions μ which do not satisfy the condition $\ulcorner (E\beta)$ $\Box(\alpha)(\alpha\mu's\equiv. \ \alpha = \beta)\urcorner$ are treated as covert descriptions or class abstracts.

Each of these three ways of handling singular terms has its virtues and vices.

Approach 2) comes close to Hilbert and Bernays's treatment of descriptions[9] and shares its main drawback, the difficulties we get in determining whether a given string of marks is a well-formed formula.[10] In most formal systems one wants an effective procedure for determining well-formedness. Hilbert and Bernays's system and approach 2) above lack such a procedure. When unlike Hilbert and Bernays one admits factual sentences into one's system, then the notion of well-formedness may even come to depend on questions of facts. This is, however, happily avoided in our case, due to the '$\Box$' in the condition, $\ulcorner (E\beta) \ \Box(\alpha) \ (\varphi \equiv. \ \alpha = \beta)$.

Another drawback of approach 2) is that the category of singular terms comes to vary not only as we pass from extensional logic to modal logic, but

---

[9] Hilbert and Bernays, *Grundlagen der Mathematik*, I, pp. 383ff.

[10] Cf., e.g., Carnap, *Meaning and Necessity*, pp. 33-34.

also as we pass from e.g., the logical modalities to other types of modalities. Thus, as we shall se in section 23, in systems of epistemic modalities the operator '□' in (2) and (2') and in the corresponding conditions in 2) must be replaced by a 'V' ('it is verified that' or 'it is known that').

Approach 3), which follows Russellian lines, does not share the former of these two defects. On approach 3) we may easily get a simple, effective procedure for determining well-formedness. The difficulty relating to our category of singular terms will, however, remain with us if we decide to regard as covert descriptions *only* those name-like expressions μ which fail to satisfy $\ulcorner(E\beta)\ \Box(\alpha)(\alpha\mu\text{'s}\equiv.\ \alpha=\beta)\urcorner$. This difficulty may, however, be overcome if we treat *all* name-like expressions, also those which satisfy this condition, as covert descriptions.

Another difficulty relating to approach 3) is that our laws of substitution and universal instantiation for descriptions come to vary as we pass from an extensional logic to modal logic, and from one type of modalities to another.

The main drawback of approach 3) seems, however, to be that we get difficulties with the universal substitutivity of identity. Thus, e.g., 'the Morning Star = the Evening Star' may be true and yet 'the Morning Star' not universally substitutable for 'the Evening Star'. These difficulties are best avoided e.g., by saying that an '=' flanked by a description is not an identity sign. It should, however, be noticed that when such a statement is expanded, an ordinary identity sign is used in the expansion. Thus, e.g., $\ulcorner(\imath\alpha)\ \varphi=\beta\urcorner$, where the '=' is not an identity sign, goes into $\ulcorner(\alpha)(\varphi\equiv.\ \alpha=\beta)\urcorner$ where the '=' is an identity sign.

This difficulty concerning the substitutivity of identity could be avoided if we based our theory of descriptions on a set of contextual definitions different from that of Russell. Instead of like Russell letting, e.g., 'G(ıx)Fx' expand into '(Ey)((x)(Fx≡.x=y).Gy)', we could let it expand into, '(Ey)(□(x)(Fx≡.x=y).Gy)'. Then any two expressions, μ and ν for which $\ulcorner\mu=\nu\urcorner$ is true could be substituted for each other in all contexts. On the other hand, a statement like 'the Morning Star = the Evening Star' would then be false.

Approach 1) avoids all the difficulties mentioned above. Its drawbacks for practical purposes are, however, obvious: The system is cumbersome, lacking the convenience afforded by a theory of descriptions and class abstraction. However, nothing prevents one who chooses alternative 1) from developing a system of shorthand notations and proofs, utilizing descriptions and class abstracts as one would do on approach 3). This shorthand notation would not be a part of the formal system itself, but would only be an adjunct to it, added in order to facilitate the use of the system.

Approach 1), supplemented by such a shorthand notation, seems to be the most recommendable approach. It combines the theoretical virtues of approach 1) with the practical advantages of approach 3).

18. Substitutivity of Identity and Other Types of Inference Turning on Singular Terms

While inference in sentential modal logic does not turn on singular terms, the advent of quantification theory and identity theory introduces three types of inference turning on singular terms, viz., universal instantiation, existential generalization, and substitution.

In the preceding section we found that in modal logic an expression μ can be expected to behave like a singular term if and only if it satisfies the condition,

$$(1) \qquad \ulcorner \Box_{(1)} \ldots \Box_{(m)} \, (E\beta) \Box_{(1)} \ldots \Box_{(n)} \, (\alpha)(\alpha\mu's \equiv . \alpha = \beta) \urcorner$$

for all natural numbers m and n. (The condition stated in the outline of approach 2) of the preceding section.) A description $\ulcorner (\imath\alpha)\varphi \urcorner$, in particular, can be expected to behave like a singular term if and only if

$$(2) \qquad \ulcorner \Box_{(1)} \ldots \Box_{(m)} \, (E\beta) \Box_{(1)} \ldots \Box_{(n)} \, (\alpha)(\varphi \equiv . \alpha = \beta) \urcorner$$

is satisfied ((2') of the preceding section).

Two other views on these three types of inference in modal logic have, however, been propounded. One is that of Smullyan, who, as mention in note 7, chapter 2, holds that all three types of inference work in modal logic if descriptions are contextually defined and one is judicious in choosing the scopes of the descriptions. Another rather widespread view is that these three principles of inference, in particular substitutivity of identity, do not work and should not be expected to work in modal contexts. In this section these two views will be examined.

Smullyan, in his review of Quine's "The problem of interpreting modal logic" and in the article, "Modality and description" argues that one is free to use the full apparatus of Russellian descriptions in modal logic, and that all the laws laid down for descriptions in *Principia Mathematica* work also in modal logic. One only has to be judicious in choosing the scopes of the descriptions, and one must in particular observe that in modal contexts the scope of a description $\ulcorner(\imath\alpha)\varphi\urcorner$ matters also when the condition

(3) $\ulcorner(E\beta)(\alpha)(\varphi\equiv.\alpha=\beta)\urcorner$

is satisfied. Also, contextually defined class abstracts can be freely admitted in modal logic if one, contrary to custom in extensional logic, uses prefixes to indicate their scope.

Many others, including Miss Barcan and Fitch, have expressed that they are largely in agreement with this view.[11] But Wilson's counter-example,[12] quoted in note 7, chapter 2 above, shows that something is wrong with it. The general difficulty illustrated by Wilson's example is that if we substitute a description for a description the scope is already selected, and there is no question of judiciously choosing the scope:

$$(\imath x)(Wx)=(\imath x)(Mx)$$
$$\Box\{[(\imath x)(Wx)\ S(\imath x)(Wx) \equiv [(\imath x)(Wx)]\ S(\imath x)(Wx)\}$$
$$\therefore \Box\{[(\imath x)(Wx)\ S(\imath x)(Wx) \equiv [(\imath x)(Mx)]\ S(\imath x)(Mx)\}$$

Similar counter-examples may be constructed also for existential generalization, using, e.g., the 'H' from our example concerning two possible worlds and two objects a and b[13]:

$$\Box[(\imath x)Hx]\ ((\imath x)Hx=(\imath xHx)$$
$$\therefore (Ey)\Box\ [\imath x)Hx]\ (y \qquad =(\imath x)(Hx)$$

---

[11] Miss Barcan in her review of "Modality and description" and Fitch in his article "The problem of the Morning Star and the Evening Star," p. 138, note 4.

[12] Wilson, *The Concept of Language*, p.43.

[13] The example concerning the two possible worlds was as follows:

Possible non-actual world	Fa.~Ga.~Ha.Ka	~Fb.Gb.Hb.~Kb
Actual world	Fa.~Ga.Ha.~Ka	~Fb.Gb.~Hb.~Kb

Here the premises is true, the conclusion is false.

A counter-example similar to these two examples is in fact supplied by Smullyan himself on p. 37 of his article. Writing '$\hat{x}\,(Ax)$' for '9', '$\hat{x}\,(Bx)$' for 'the number of planets', and 'f' for 'is less than 10', Smullyan gets"

$$\hat{x}\,(Ax)=\hat{x}\,(Bx)$$
$$\Box([\hat{x}\,(Ax)]\;f\;\hat{x}\,(Ax))$$
$$\Box([\hat{x}\,(Bx)]\;f\;\hat{x}\,(Bx))$$

(the scope symbols are put in by me for emphasis. Smullyan leaves them out on the convention that the scope of the abstract is understood to be the shortest formula containing the abstract unless otherwise noted.)

Smullyan remarks that here the conclusion cannot be deduced from the premises. But he adds that "if in the second premises and the conclusion the class abstract is given maximum scope, the conclusion would be valid." He then concludes from this example that "the modal paradoxes arise out of neglect of the circumstance that in modal contexts the scopes of incomplete symbols, such as abstracts or descriptions, affect the truth value of those contexts." (p. 37) Smullyan's defense against the other two counter-examples mentioned above would probably be similar: He would observe, for example, in the case of Wilson's example that if in the second premises and the conclusion the descriptions are given the maximum scope, one would get the following valid argument:

$$(\imath x)(Wx)=(\imath x)(\Diamond x)$$
$$[(\imath x)(Wx)]\;[(\imath x)(Wx)]\;\Box\,\{S(\imath x)(Wx)\equiv S(\imath x)(Wx)\}$$
$$\therefore\;[(\imath x)(Wx)]\;[(\imath x)(\Diamond x)]\;\Box\,\{S(\imath x)(Wx)\equiv S(\imath x)(\Diamond x)\}$$

That this weaker conclusion is obtainable was, however, observed by Wilson himself. And Smullyan's defense is a poor one. For the claim of Smullyan's paper can then not be that "the unrestricted use of modal operators in connection with statements *and* matrices embedded in the framework of a logical system such as *Principia Mathematica* does not involve a violation of Leibniz's principle."[14] His claim dwindles apparently to the far weaker: In a system of the type just described, any argument utilizing Leibniz's principle, however fallacious it may be, can be turned into a correct argument if we judiciously

---

[14] Smullyan, "Modality and description," p.34.

change the scopes in the premises and the conclusion (or judiciously choose them, if not all scopes in the argument have been selected beforehand).

In Smullyan's article there is also another claim which seems untenable. Discussing class abstracts, Smullyan proposes as axioms of abstraction and extensionality respectively:

$$(4) \qquad \Box(E\alpha)(\varphi x \equiv_x x\varepsilon\alpha)$$

and

$$(5) \qquad \Box(x\varepsilon\alpha \equiv_x x\varepsilon\beta. \supset \alpha = \beta)$$

observing that if in place of (4) he had stipulated

$$(6) \qquad (E\alpha)\Box(\varphi x \equiv_x x\varepsilon\alpha)$$

paradoxes would arise, among that that no true formal equivalences are contingent (Smullyan, p.36-37).

Now (6) corresponds to the condition for descriptions we arrived at in the previous section, viz.,

$$(7) \qquad \ulcorner(E\beta)\Box (\alpha)(\varphi \equiv .\alpha = \beta)\urcorner$$

Condition (4) is weaker; the corresponding condition for descriptions,

$$(8) \qquad \ulcorner\Box(E\beta)(\alpha)(\varphi \equiv .\alpha = \beta)\urcorner$$

would be satisfied by expressions like, '$(\imath x)Hx$', where 'H' is the 'H' of our example concerning the two objects a and b in the two possible worlds (note 13 above). In the preceding section we found that unless a description satisfies (7) it cannot be expected to behave like a singular term. And examples showing that (8) is too weak a condition are easily forthcoming. Thus in our example concerning our two possible worlds, the premises of the following argument are both true, and the conclusion false:

$$(\imath x)Hx = (\imath x)Fx$$
$$\Box[(\imath x)Fx] \, F \, (\imath x)Fx$$
$$\therefore \Box[(\imath x)Hx] \, F \, (\imath x) \, Hx$$

If in the second premise and the conclusion large scope had been used, the inference would have been valid. This incidentally shows that scope matters also when the condition (8) is satisfied and hence given an independent proof that there are descriptions which satisfy (8) but do not behave as singular terms. (A description behaves as a singular term only if no conventions regarding scope are needed.)

One may feel that our model concerning two possible worlds is a rather artificial one, especially since in it 'H' is true of one and only one object in each possible world, but of different objects. Such qualms do not affect the validity of the above refutation of Smullyan, since our model with two possible worlds satisfies all the axioms and rules of even the strongest system of modal logic proposed, viz., S5, and still the premises in our example come out true and the conclusion false. There seem, anyway, to be predicates which behave like 'H' also in less artificial models. The predicate, 'is the number of planets' seems to be a case. For although this predicate is true of one number, 9, in our actual world and of other numbers in other possible worlds, it is presumably true of one and only one natural number in each possible world (0 being a natural number). The number of material objects of any specified type would apparently be equally good candidates for our 'H'.

Miss Barcan, in her review of Smullyan's article, and Fitch, in "The problem of the Morning Star and the Evening Star," point out that if Smullyan replaces his condition (5) with

$$(9) \qquad \ulcorner \Box(x\varepsilon\alpha \equiv_x x\varepsilon\beta) \supset. \ \alpha=\beta \urcorner$$

(6) may take the place of (4) without paradoxes.[15] Miss Barcan seems to prefer this alternative, since (5) easily permits her to derive that if two classes have the same members then they necessarily have the same members.[16] But Fitch apparently prefers the first alternative, stating on p.140 of his paper that, "in modal logic the condition for the interchangeability of scopes of '$(\iota x)fx$' is '$\Box E!(\iota x)fx$'."

Smullyan, Miss Barcan, and Fitch hoped to overcome the difficulties connected with substitutivity of identity in modal contexts by using contextually defined descriptions and manipulating the scope symbols. Another, apparently

---

[15] Miss Barcan's review of "Modality and description," p.150, Fitch, "The problem of the Morning Star and the Evening Star," p.138, footnote 4.

[16] Miss Barcan's review of "Modality and description," p. 43.

less complicated way out would be to abandon Leibniz's principle and rest contented with a principle of restricted substitutivity of identity, e.g., substitutivity in extensional contexts only. Carnap's treatment of what he calls "identity of extension" seems to be an attempt in this direction.

But the reasons against restricting the substitutivity of identity are many. One is standard usage. A relation in a deductive system is usually called an identity relation if and only if it is reflexive and *substitutive* with respect to all contexts expressible in the system.[17]

Another, related reason, is our intuitions concerning identity, the feeling many have, that, "if identity does not mean universal interchangeability, then I do not really understand identity at all."[18]

A third reason may be had from the discussion of identity of individuals in sections 12 and 13. We there found that unless we took the identity relation to be universally substitutive, we were apparently unable to derive:

$$(10) \qquad (x)\,(y)\,(x{=}y \,.\, \supset \Box(x{=}y))$$

let alone the stronger,

$$(10') \qquad \Box_{(1)}\,\Box_{(2)}\ldots\Box_{(m)}\,(x)(y)(x{=}y \,.\, \supset\Box_{(1)}\,\Box_{(2)}\ldots\Box_{(n)}\,(x{=}y))$$

And in section 12 we found (10) and (10') to be true in any interpreted system of modal logic, so that according to Corollary 1.3, and system of quantified modal logic with identity in which (10) is not derivable is semantically incomplete. Also, logical considerations concerning completeness therefore demand universal substitutivity of identity.

It should be noticed that the reasoning leading up to Thesis 1 and its corollaries, among them Corollary 1.3, at no point presupposed anything about the substitutivity of identity. What concerned us was exclusively the interpretation of the quantifiers.

Therefore, observing that, e.g., the expressions 'the Morning Star' and 'the Evening Star' cannot be substituted for each other in all modal contexts, let us not jump to the conclusion that the identity relation is not universally

---

[17] See, e.g., A. A. Fraenkel, "The relation of equality in deductive systems" and Quine's review of this article in *The Journal of Symbolic Logic*, 14 (1949), p.130.

[18] Wilson, *The Concept of Language*, p.39.

substitutive. Let us rather cease to consider expressions like this as names, as indicated in the preceding section. Or, if we insist that they are names, let us then admit, with Frege, that in modal contexts we talk not about material objects, but about intensions: concepts, attributes, and propositions.

Chapter Six

Examination of the Difficulties

19. Examination of the Difficulties Surveyed in Section 10.

With the results from the preceding discussion at hand, let us now examine the difficulties relating to quantification into modal contexts which were surveyed in section 10. We will identify the arguments by the numbers (1-9) which we used in section 10.

1. The statement

(1) $(Ex)\Box(x>7)$

was found to be disturbing, because apparently 9, but not the number of planets is one of the numbers whose existence is affirmed in (1). If, however, we cease to regard the expression 'the number of planets' as a name, then we may well accept the fact that,

(2) $\Box(9>7)$

is true and nevertheless,

(3) $\Box(\text{the number of planets} > 7)$

is false. For on neither of the three approaches in section 17 would

(4)   9 = the number of planets

be an identity sentence. If we take approach 3), (4) would be well-formed and true, but the '=' would not be an identity sign. If we choose approaches 1) or 2), (4) would not be well-formed; we should rather write:

(4')   (Ey)((x) (the planets are x in number $\equiv$.x=y) .y=9)

and (3) would correspondingly become,

(3') $\Box$(Ey)((x)(the planets are x in number $\equiv$.x=y) .y>7)

(Since as mentioned in the preceding section, there is in each possible world exactly one natural number of which 'the planets are x in number' is true, the clause '(x)(the planets are x in number $\equiv$.x=y)' in (4') and (3') could be shortened to 'the planets are y in number'.)

2. The observation that if substitutivity of identity breaks down in modal contexts, then we cannot quantify into them, has been found to be correct. In section 12 we saw that in any interpreted system of quantified modal logic with identity,

(5) $\Box_{(1)} \ldots \Box_{(m)}$ (x)(y)(x=y . $\supset \Box_{(1)} \ldots \Box_{(n)}$ (x=y))

is true. In section 17, we found that (5) would be satisfied if and only if we require that all of our singular terms keep their references in all possible worlds. But this requirement also removes the cause of the apparent breakdown of substitutivity of identity, which was that two singular terms could be co-referential in one possible world but not in another. If we want to quantify into modal logic contexts we must hence require that substitutivity of identity holds in them. Something therefore has to be done with statements like,

(6) $\Box$ (the number of planets >7))

which in section 7 led us to conclude tentatively that substitutivity of identity breaks down in modal contexts.

Church's proposal, that in modal contexts 'the number of planets' does not refer to the object referred to by '9', is a way out. We leave our stock of singular terms untouched, and individuate the entities of our universe of discourse finely enough to secure that all our singular terms have one and only one reference which they may keep in all possible worlds. That is, we quantify over intensions.

Another way out is sketched in section 17: we quantify over extensions, but restrict our stock of definite singular terms, either by eliminating them completely (approach 1)) or by keeping only those of them which refer to the same extensional object in all possible worlds (approaches 2) and 3)).

3. The unusual character which the development of an intensional language adequate to general purposes would otherwise have to assume, is an argument in favor of restricting the stock of singular terms rather than extruding extensional objects from the range of our variables. The sentence,

(7) The number of planets is a power of three

may be kept unchanged if we choose approach 3), while if we take approaches 1) or 2) it goes, for example, into,

(7') The planets number a power of three

The sentence,

(8) The wives of two of the directors are deaf

may remain unchanged on all three approaches.

4. Quine's argument relating to congruents shows that if 'Morning Star' and 'Evening Star' are among our singular terms in quantified modal logic with identity, our ontology is intensional.

In order to avoid an intensional ontology one must cease to regard the two expressions as singular terms and rather treat them, e.g., as 'Hesperus' and 'Phosphorus' were treated in section 17. Quine's 'C' is then no longer the relation of identity, but corresponds rather to a relation of co-extensiveness between predicates true of one and only one object.

5. Although,

(9) $(x)(y) (x=y . \supset \Box (x=y))$

is provable in Miss Barcan's modal calculi, the values of her variables do not therefore have to be intensions. As we saw in section 17, (9) does not require extensional objects to be extruded from one's universe of discourse provided one's stock of singular terms is appropriately restricted.

It goes without saying that even with an appropriately restricted stock of singular terms, we are free to commit ourselves to an intensional ontology if we wish. We may, e.g., affirm statements like '(x)(x is an intensional object)'. In this respect, modal logic is similar to extensional logic. The point of section 17 and the observation above concerning Miss Barcan's systems is that quantification into modal contexts and results like (9) above do not compel us to extrude extensional objects from our universe of discourse.

6. It is right that in modal logic the universal closure of

$$(10) \qquad \Box \ (x=\sqrt{x} + \sqrt{x} + \sqrt{x} \neq \sqrt{x} \, . \supset . \ x>7)$$

is true, while the universal closure of

$$(11) \qquad \Box(\text{there are exactly x planets} \supset . \ x>7)$$

is false. This shows, as was mentioned in sections 6 and 7, that co-extensional open sentences cannot be interchanged in modal contexts *salva veritate* as they can in extensional contexts. It also indicates that the modal operator '□' is rather like the expression 'is valid' or 'is logically true' except that the latter two attach to names of sentences, while '□' attaches to sentences (section 6).

But that co-extensional general terms be interchangeable in a context is not required in order that we shall be able to quantify into the context.

If we want to quantify into modal contexts, we must, however, require that if we admit open sentences like,

$$(12) \qquad \Box(x>7)$$

or

$$(13) \qquad \Box(\text{if there is life on the Evening Star then there is life on x})$$

into our system of modal logic, fulfillment of them by the objects over which we quantify must make sense.

If, as indicated in section 17, we expurgate from our stock of singular terms those expressions μ which do not satisfy the requirement $\ulcorner(E\beta)\ \Box(\alpha)(\alpha\mu's\equiv.\ \alpha = \beta)\urcorner$ then 'the number of planets', 'the Evening Star' and 'the Morning Star' all disappear from our vocabulary of singular terms.

On all the three approaches outlined in section 17, the open sentence (12) above may be kept unchanged and becomes true of, e.g., the number 9, false of, e.g., 5.

The open sentence (13) may be kept unchanged when we use approach 3), but becomes unacceptable if we take approaches 1) or 2). Instead we might, for example, write,

> (13') $\Box$ (there is one and only one star which shines conspicuously bright in the evenings, and if there is life on this star then there is life on x).

7. If, in order to overcome the difficulties confronting quantified modal logic, one requires that any two conditions which uniquely determine an object in our universe of discourse be analytically equivalent, then, as Quine points out in *From a Logical Point of View*,

> (14) $(x)(y)(x=y. \supset \Box\ (x=y))$

will be true.

Since, however, as we saw in section 12, we have to accept (14) anyway, and since Quine in *Word and Object* derives more disastrous consequences from the above requirement, let us postpone the examination of it to part 9 of this section.

8. The three results at which Quine arrives in his discussion of quantified modal logic in "Three grades of modal involvement" have all been arrived at again in this thesis.

In section 12 we found that

> (14) $(x)(y)(x=y. \supset \Box\ (x=y))$

and even the stronger,

> (15) $\Box_{(1)} \ldots \Box_{(m)}\ (x=y\ .\supset \Box_{(1)}\ \ldots\ \Box_{(n)}\ (x=y))$

is true in every interpreted system of quantified modal logic with identity. (14) and (15) were true, not only because standard usage requires an identity relation to be universally substitutive, but because interpretation of the quantifiers was found to be impossible unless (15) and (14) are true.

Quine's remark in "Three grades of modal involvement" (p.81) that interference in the contextual definition of singular terms even when their objects exist might be a way out of the difficulties, has been followed up in this thesis. The three approaches sketched at the end of section 17 are all based on manipulation with the singular terms.

Thirdly, Aristotelian essentialism is unavoidable in quantified modal logic. But if the modal operator '□' itself makes sense, then, as we saw in part 6 of this section, open sentences with an '□' prefixed make sense too if we restrict our stock of singular terms appropriately.

To make sense of Aristotelian essentialism and to make sense of open sentences with an '□' prefixed are one and the same problem, and a solution to the one is a solution to the other. Much of this thesis, especially section 17, is therefore indirectly a discussion of Aristotelian essentialism.

9. In *Word and Object*, Quine finds that if conditions which uniquely determine the same object in our universe of discourse are required to be analytically equivalent, then modal distinctions collapse.

In section 16 modal distinctions were found to collapse if descriptions (or class abstracts) are treated as names and in addition some other rather general conditions are satisfied (Theorems XVI and XVII). We therefore had to introduce a rather severe restriction on the type of descriptions we could permit to be treated as names: $\ulcorner(E\beta) \; \Box(\alpha)(\alpha\mu's\equiv. \; \alpha = \beta)\urcorner$ had to be true in order that $\ulcorner(\int\alpha)\varphi\urcorner$ could be treated as a name.

If the above restrictions are adhered to, and if our modal system is based on Gödel's basic system or a stronger system, then one may easily prove that any two conditions *which are allowed to occur in descriptions which are treated as names* are analytically equivalent if they uniquely determine the same object. For let 'Fx' and 'Gx' represent two such conditions. Then we have:

*(1')	$Fx\equiv.x=y:Gx\equiv.x=y$	
*(2')	$Fx\equiv Gx$	(1')
(3')	$Fx\equiv.x=y:Gx\equiv.x=y: \supset.Fx\equiv Gx$	*(2')
(4')	$\Box \; (Fx\equiv.x=y:Gx\equiv.x=y: \supset.Fx\equiv Gx)$	(3') RL
*(5')	$(Ey)(\Box(x)(Fx \equiv. \; x=y). \; \Box(x)(Gx \equiv.x=y))$	by hypothesis
*(6')	$\Box(x)(Fx \equiv. \; x=y). \; \Box(x)(Gx \equiv.x=y))$	(5') y

*(7')	$(x)\square(Fx \equiv. x=y). (x)\square(Gx \equiv.x=y))$	(6') by (2) of section 15
*(8')	$(x)\square(Fx \equiv. x=y:Gx \equiv.x=y)$	(7') by (5) of section 16
*(9')	$\square(Fx \equiv. x=y:Gx \equiv.x=y)$	(8')
*(10')	$\square (Fx\equiv.x=y:Gx\equiv.x=y: \supset.Fx\equiv Gx) \supset.$ $\square(Fx \equiv. x=y:Gx \equiv.x=y)) \supset \square(Fx\equiv Gx)$	Gödel's axiom A.2
*(11')	$\square (Fx\equiv Gx)$	(4') (9') and (10')
*(12')	$(x) \square (Fx\equiv Gx)$	(11') x

In fact, any such condition turns out to be analytically or necessarily true of the object it determines. And if Brouwer's axiom $\ulcorner\varphi\square\Diamond\varphi\urcorner$ is a theorem of the system, then one may also prove that any such condition is necessarily false of all other objects in our universe of discourse.[1]

---

[1] For let 'Fx' represent such a condition. Then one has:

(1)	$y=y$	
(2)	$\square(y=y)$	(1) RL
(3)	$Fy\equiv.y=y: \supset:y=y. \supset Fy$	
(4)	$\square (Fy\equiv.y=y: \supset:y=y. \supset Fy)$	(3) RL
*(5)	$(Ey)\square(x)(Fx \equiv. x=y)$	by hypothesis
*(6)	$\square(x)(Fx \equiv. x=y)$	(5) y
*(7)	$(x)\square(Fx \equiv. x=y)$	(6) by (2) of section 15
*(8)	$\square(Fy \equiv. y=y)$	
*(9)	$\square(Fy \equiv. y=y:\supset:y=y. \supset Fy)\supset.$ $\square(Fy \equiv. y=y) \supset\square(y=y. \supset Fy)$	by Gödel's axiom A.2
*(10)	$\square (y=y . \supset Fy) \supset. \square(y=y) \supset\square Fy$	by Gödel's axiom A.2
*(11)	$\square Fy$	(2) (4) (8) (9) and (10)
*(12)	$(Ex) \square Fx$	(11)

The requirement that two conditions 'Fx' and 'G'x' which uniquely determine the same object in our universe of discourse are analytically equivalent, is as we saw in section 17 (approach 3)), all that is needed in order that the descriptions '(∫x)(Fx)' and '(∫x)(Gx)' may be substituted for each other in modal contexts. But as observed in section 17, if the description '(∫x)(Fx)' shall behave in al respects like a singular term, such that it, e.g., may be used to instantiate universal quantifications, then 'Fx' has to satisfy the stronger requirement '(Ey) □(x)(Fx≡.x=y)'

Open sentences which do not occur in descriptions which are substituted for each other or in other ways treated like proper names do not, however, have to be analytically equivalent even if they uniquely determine the same object.

We may therefore conclude that if open sentences occur in descriptions which are substituted for each other, they have to be analytically equivalent.

But open sentences which do not occur in such descriptions do not have to be analytically equivalent, even if they uniquely determine the same object. For as mentioned in part 6 of this section, neither quantification theory nor identity theory requires co-extensional sentences to be interchangeable. What we may

---

And if Brouwer's axiom $\ulcorner \varphi \supset \Box \Diamond \varphi \urcorner$ is a theorem of the system, then one also has:

(1)	Fx≡.x=y: ⊃ : x≠y. ⊃~Fx	
(2)	□ (Fx≡.x=y: ⊃ : x≠y. ⊃~Fx)	(2) RL
*(3)	(Ey) □(x)(Fx≡.x=y)	by hypothesis
*(4)	□(x)(Fx≡.x=y)	(3) y
*(5)	(x)□ (Fx≡.x=y)	(4) by (2) of section 15
*(6)	□ (Fx≡.x=y)	(5)
**(7)	x≠y	
**(8)	x≠y. ⊃ □(x≠y	Theorem XV (section 15)
**(9)	□ (Fx≡.x=y: ⊃ : x≠y. ⊃~Fx) ⊃. □ (Fx≡.x=y) ⊃ □ (x≠y. ⊃~Fx)	by Gödel's axiom A.2
**(10)	□ (x≠y. ⊃ ~Fx) ⊃.□(x≠y) ⊃□~Fx	by Gödel's axiom A.2
**(11)	□~Fx	(2) (6) (7) (8) (9) (10)
*(12)	x≠y. ⊃□~Fx	*(11)
*(13)	(x)(x≠y. ⊃□~Fx)	(12) x
*(14)	(Ey)(x)(x≠y. ⊃□~Fx)	(13)

learn from the argument in *Word and Object* is therefore that in quantified modal logic one cannot use a standard theory of descriptions. If one does, modal distinctions collapse.

20. Other Difficulties in Quantified Modal Logic.

Although quantified modal logic can be made to work without a collapse of modal distinctions and without extrusion of extensional objects from our universe of discourse, there is a feature of quantified modal logic seems rather undesirable: interpretation of the quantifiers requires that every object in the actual world exists in every possible world, i.e., *every object exists necessarily*. We arrived at this result in section 12 (result (9)). Indeed, we found that,

$$(1) \qquad \Box_{(1)} \ldots \Box_{(m)} \, (x)(Ey) \, \Box_{(1)} \ldots \Box_{(n)}(x=y)$$

had to be true in every interpreted system of quantified modal logic with identity (Corollary 1.2).

Formula (1) follows by quantification theory from,

$$(2) \qquad \Box_{(1)} \ldots \Box_{(m)} \, (x)(y) \, (x=y \, . \, \supset \Box_{(1)} \ldots \Box_{(n)}(x=y))$$

As we saw in section 13, (2) required for its proof that the identity relation be universally substitutive. It should, however, be noticed that (1), unlike (2), is provable also in systems in which the identity relation is not universally substitutive. A proof of (1) (with m=0, n=1) might, e.g., go as follows:

(1')	$x=x$	axiom of identity
(2')	$\Box \, (x=x)$	(1') RL
(3')	$(Ey) \, \Box(x=y)$	(2')
(4')	$(x)(Ey) \, \Box(x=y)$	(3') x

Repeated applications of the rule RL to line (2') and to the end result (1) for arbitrary *m* and *n*. Hence, not all seemingly counter-intuitive results in modal logic can be blamed on the unrestricted substitutive of identity. (Besides, as we saw in section 12, restrictions on the substitutivity of identity are of no avail, since the formulae which such restrictions prevent us from deriving nevertheless remain true under all interpretations.)

That (1) is counter-intuitive seems to be beyond doubt. To be sure, not all name-like expressions have to refer to some object in every possible world. The expression 'Hersperus' does not, for example, need an object corresponding to it in all possible worlds. But the object which in our actual world "Hesperizes" has to exist in all possible worlds. And, although such a view might be comforting, turning as it does, everybody and everything into necessarily existing half-gods, it is hardly desired by modal logicians.

In view of this result it might perhaps be preferable after all to extrude material objects from our universe of discourse and keep only intensions. For individual concepts, unlike individuals, might well be granted necessary existence.

Another way out, which permits us to retain the advantages of variables taking extensional objects as values, would be to treat *existence as a predicate*, which could be affirmed or denied of the members of our universe of discourse. We would then not quantify over existing objects exclusively, but also over non-existing objects, a kind of actual, but unactualized possibles. There are certainly difficulties in connection with such an approach. First the difficulty of distinguishing two sense of 'exists' or 'is': that of the existential quantifier and that of the predicate which may be denied of some members of our universe of discourse. In addition, there would be rather many different kings of "possibles" on such an approach, some of them actual possibles, others merely possible possibles.

Whichever way out we choose, the interpretation of the quantifiers requires (1) to stay with us. To put up with (1) and the difficulties which flow from it, seems to be the main price one has to pay if one wants to quantify into modal contexts.

Chapter Seven

Other Types of Opaque Contexts

21. Causal Modalities. Counterfactuals.

In this chapter an attempt will be made to carry the results reached in the preceding chapters over to other types of opaque contexts. Our aim will not be to axiomatize these other types of discourse or to compare various proposed axiomatizations and discuss their merits and shortcomings. We will only briefly outline how these other contexts apparently can be assimilated to the logical modalities, mention some differences between these contexts and the logical modalities, and then seek to carry the results from the earlier chapters over to these new contexts. What will concern us is the referential and extensional *opacity* of these contexts, and in particular whether one may quantify into them.

Let us start with the contexts which seem most similar to the logical modalities.

The *natural necessities* may apparently be interpreted along lines exactly parallel to those which we followed for the logical modalities in section 11.

Instead of talking about worlds which are logically possible, we talk about worlds which are physically possible. Just as the former worlds were those which were compatible with the laws of logic, the latter worlds will be those which are compatible with the laws of nature. The latter worlds will obviously be compatible with the laws of logic, but the former worlds will not always be compatible with the laws of nature. Writing 'C' for 'it is physically, or causally necessary that,' we will hence have:

$$\vdash \ulcorner \Box\varphi \supset C\varphi \urcorner$$

while the converse will not always hold.[1] Further, we obviously have,

$$\ulcorner C\varphi \supset \varphi \urcorner$$

In order that the causal modalities shall not collapse, contexts governed by a 'C' have to be non-extensional. But, in view of the discussion in the preceding chapters, we should nevertheless be able to quantify into them if we are willing to restrict our stock of singular terms and to accept difficulties similar to those discussed in the preceding section.

By reasoning exactly parallel to that of section 12, it is seen that in order to quantify into contexts of natural necessity, one must require that,

$$(1) \qquad (x)\,(y)\,(x{=}y \,.\, \supset C(x{=}y))$$

If iterated modalities are permitted, additional 'C''s will be needed in (1). Any system which permits quantification into natural necessity contexts will be semantically incomplete unless (1) is derivable in it.

Due to (1), descriptions and class abstracts may not in general be treated as names in the causal modalities; we have to restrict our stock of singular terms as indicated in Chapter Five, treating, e.g., a description $\ulcorner(\imath\alpha)\varphi\urcorner$ as a singular term only if it satisfies the condition,

$$(E\beta)C(\alpha)(\varphi \equiv .\, \alpha{=}\beta)$$

From (1) we may derive,

---

[1] In ordinary usage, 'causally, or physically necessary' seems to have a slightly more narrow sense. One would thus perhaps not ordinarily say that it is physically, or causally necessary that 2+2=4. If this should be the case, $\ulcorner\varphi$ is causally necessary$\urcorner$ would perhaps correspond to $\ulcorner C\varphi.\sim \Box\varphi\urcorner$ rather than to $\ulcorner C\varphi\urcorner$. This does not, however, affect our discussion; we might continue to use 'C' in the broader sense and paraphrase $\ulcorner\varphi$ is causally necessary$\urcorner$ as $\ulcorner C\varphi \,.\, \Box\varphi\urcorner$.

(2)   (x)(Ey)C(x=y)

That is, every object in our actual world has to exist in every world which is compatible with the laws of nature. The difficulties discussed in the preceding section therefore recur in the logic of the natural necessities.[2]

If, however, we are willing to accept these restrictions and difficulties, and if we can make sense of the operator 'C' itself, then we do apparently have a way of handling natural law statements and contrary-to-fact conditions. 'All emeralds are green' could be paraphrased into:

$$(x)C(x \text{ is an emerald} \supset x \text{ is green})$$

'If *a* were an emerald, it would have been green' would similarly go into,

$$\sim(a \text{ is an emerald}).(x)C(x \text{ is an emerald} \supset x \text{ is green})$$

We would still have difficulties with many types of counterfactual conditionals like Quine's example,

> If Bizet and Verdi had been compatriots, Bizet would have been Italian

(*Methods of Logic*, p. 15). But these difficulties apparently do not relate to the opacity of contexts governed by the operator 'C', nor to the quantification into such contexts, but relate rather to the sense of the operator 'C' itself.

---

[2] The only proposed system of quantified causal logic is that of Burks, presented in "The logic of causal propositions" (1951). This system does not contain a theory of identity. It does, however, contain a full set of axioms and rules for quantification theory, and has also the rule of inference RL (If $\vdash \varphi$, then $\vdash \ulcorner \Box \varphi \urcorner$) (Burks's rule III, p. 381 of his article). Since $\ulcorner \Box \varphi \supset C \varphi \urcorner$ is an axiom of Burks's (his axiom (6), p. 381) and modus ponens is one of his rules (his rule I, p. 381), it follows that when a set of axioms for identity theory is added to his system, (1) may be proven in his system along lines exactly parallel to those we followed in order to prove the corresponding theorem for the logical modalities in the beginning of section 13. And (2) may similarly be proven along the lines by which '(x)(Ey)$\Box$(x=y)' was proven in the beginning of section 20.

22. *Deontic Modalities.*

The key expressions of the deontic modalities are 'it is obligatory that' and 'it is permitted that'. Seeking to interpret the deontic modalities along the lines of section 11, we might tentatively take the permissible to be that which is true in an ethically "ideal" world, i.e., a world which is as it ought to be, a world compatible with the principles of ethics. The obligatory would correspondingly be that which is true in every such ideal world. Writing 'O' for 'it is obligatory that' and 'P' for 'it is permitted that', we can then expect the operators 'O' and 'P' to behave approximately like the operators '□' and '◇', respectively, of the logical modalities. Thus, e.g., we have,

$$\ulcorner O\varphi \equiv {\sim} P {\sim} \varphi \urcorner$$

But since the actual world is a possible world, but presumably not an ideal world, we can also expect conspicuous differences between 'O' and '□', and 'P' and '◇'.

Thus we have $\ulcorner \Box \varphi \supset \varphi \urcorner$ but not in general $\ulcorner O\varphi \supset \varphi \urcorner$, and $\ulcorner \varphi \supset \Diamond \varphi \urcorner$ but not in general $\ulcorner \varphi \supset P\varphi \urcorner$. That is, not all that is obligatory is done, and not all that is done is permitted. Instead of $\ulcorner O\varphi \supset \varphi \urcorner$ we might perhaps use,

$$(1) \qquad \ulcorner OO\varphi \supset O\varphi \urcorner$$

or

$$(2) \qquad \ulcorner O(O\varphi \supset \varphi) \urcorner$$

as a principle of deontic logic. And likewise instead of $\ulcorner \varphi \supset P\varphi \urcorner$ we might consider using,

$$\ulcorner P\varphi \supset PP\varphi \urcorner$$

or

$$\ulcorner O(\varphi \supset P\varphi) \urcorner$$

In spite of these and other differences between the logical and deontic modalities, it seems, however, possible to interpret the deontic modalities in terms of a set of ideal worlds supplemented by our actual world, as indicated above. And if so, we should expect that many of the results reached in earlier chapters may be transferred to the deontic modalities.

In particular, it should be possible to quantify into deontic contexts from outside if we accept the restrictions at which we arrived in the earlier chapters. We must, e.g., according to section 12 require that,

$$(3) \qquad (x)(y)(x=y . \supset O(x=y))$$

Since iterated modalities seem to be required in deontic logic, as illustrated by (1) and (2) above, we must in fact require that the stronger,

$$(4) \qquad O_{(1)} \ldots O_{(m)} (x=y . \supset O_{(1)} \ldots O_{(n)} (x=y))$$

be true in any interpreted system of quantified deontic logic.

In section 13 we proved what corresponds to (4) for the logical modalities without appealing to the principle $\ulcorner \Box \varphi \supset \varphi \urcorner$, and we should hence expect that (4) may be provable in systems of deontic logic although these systems lack the principle, $\ulcorner O\varphi \supset \varphi \urcorner$.[3]

---

[3] The only proposed system of quantified deontic logic is that of Hintikka, which is outlined in "Quantifiers in deontic logic" (1957). The system provides a general method of testing the consistency of quantificational formulae, modal and non-modal. Starting out with a set of formulae to be tested, we derive sequents from these according to certain rules. If and only if the formula or set of formulae to be tested is inconsistent, some "modal set" of sequents will contain a formula together with its negation.

Hintikka's system lacks identity. In order to see how the formulae (3), (4), and (5) above fare in Hintikka's system, we may, however, add some rules governing identity to the rules which Hintikka gives on p. 14 of his paper. We might, e.g., add the two rules,

(E.8)　　If $\mu\varepsilon S'$, but not 'a=a'$\varepsilon\mu$ for some free variable 'a' occurring in the formulae of μ, then we may adjoin 'a=a' to μ.

(E.9)   If μεS', fεμ and 'a=b'εμ for some free variable 'a' of f, but not
f(b/a)εμ, then we may adjoin 'f(b/a)' to μ.

A proof of (3) might then run as follows in Hintikka's system:

Since the negation of (3) is '(Ex)(Ey)(x=y.P~(x=y))', μ comes to contain
the following formulae:

(1')	(Ex)(Ey)(x=y.P~(x=y))	
(2')	(Ey)(a=y.P~(a=y))	(1') by (E.3)
(3')	a=b.P~(a=b)	(2') by (E.3)
(4')	a=b	(3') by (E.11)
(5')	P~(a=b)	(3') by (E.12)
(6')	P~(a=a)	(4') (5') by (E.9)

μ* comes to contain,

(1")	~(a=a)	(6') by (E.5)
(2")	a=a	by (E.8)

Since the model set μ* contains a formula and its negation, the formula
from which we started out, '(Ex)(Ey)(x=y. P ~(x=y))', is inconsistent, and
'(x)(y)(x=y . ⊃ O(x=y))' is valid. A proof of (4) for greater values of *m* and *n*
can be given along similar lines.

A proof of (5) for m=0, n=1, might go as follows:

μ:
(1')	(Ex)(y)P~(x=y)	
(2')	(y)P~(a=y)	(1') by (E.3)
(3')	P~(a=a)	(2') by (E.4)

μ*:
(1")	~(a=a)	(3') by (E.5)
(2")	a=a	by (E.8)

μ* contains a formula and its negation. '(Ex)(y)P-(x=y)' is therefore inconsistent
and (5) valid. A proof of (5) for greater values of *m* and *n* can be given along
similar lines.

Reasoning parallel to that in Chapter Five shows further that in quantified deontic logic we have to restrict our stock of singular terms. Thus, e.g., a description $\ulcorner(\imath\alpha)\varphi\urcorner$ may be treated as a name only on the condition:

$$\ulcorner O_{(1)} \ldots O_{(m)} (E\beta)O_{(1)} \ldots O_{(n)}(\alpha)(\varphi \equiv .\alpha=\beta)\urcorner$$

Similar restrictions hold for class abstracts and other name-like expressions, as indicated in section 17.

Since as a consequence of (4) we have,

(5) $O_{(1)} \ldots O_{(m)} (x)(Ey)O_{(1)} \ldots O_{(n)} (x=y)$

difficulties parallel to those discussed in section 20 recur in quantified modal logic: in order to quantify into deontic contexts, one must require that every object in our universe of discourse exists in every ideal world.

It should be observed that if instead of rule (E.9) above we had used the restricted rule,

(E.9') If $\mu\varepsilon S'$, $f\varepsilon\mu$, and 'a=b'$\varepsilon\mu$ for some free variable 'a' of f, but not f(b/a)$\varepsilon\mu$, then, *if f contains no modal operator*, we may adjoin 'f(b/a)' to μ

the step from (5') to (6') in the proof of (3) would be illegitimate, and (3) would apparently not be provable in the system. The proof of (5) would, however, be unaffected by such a restriction. Such a restriction on the substitutivity of identity would make the '=' behave much like the '≡' of Carnap's system of logical modalities S_2 in *Meaning and Necessity*, and these observations concerning the derivability of (3), (4), and (5) accord well with what was proved about Carnap's system S_2 in section 13. That (5) is provable in systems of logical modalities even if substitutivity is restricted to extensional contexts was noted in note 12, chapter 3, and proven in the beginning of section 20.

23. Epistemic Modalities.

The key expression in the epistemic modalities is 'it is known that' or 'it is verified that', for which we will write 'V'. The epistemic modalities may be interpreted along the lines of section 11. Instead of talking about possible worlds, we come to talk about worlds which are compatible with our knowledge. Obviously, ⌜Vφ⌝ then comes to be interpreted as ⌜φ us true in all worlds which are compatible with our knowledge⌝. 'V' thus corresponds to the '□' in the logical modalities. Corresponding to 'possible' we get the epistemic 'not known to be false'.

Since all that is known is true, the principle,

$$⌜V\phi \supset \phi⌝$$

holds in epistemic logic. That is, our actual world is compatible with out knowledge.

The epistemic modalities therefore seem to be more closely parallel to the logical modalities than are the deontic modalities, and we can expect that the results from the earlier parts of the thesis may be transferred to the epistemic modalities.

In particular, although epistemic contexts are non-extensional, it should be possible to quantify into them if one accepts the restrictions at which we arrived in earlier chapters.

We have to require that,

$$(x)(y)(x=y. \supset V(x=y))$$

or, permitting iterated modalities,

$$(1) \qquad V_{(1)} \ldots V_{(m)} (x)(y)(x=y. \supset V_{(1)} \ldots V_{(n)} (x=y))$$

Further must we restrict our stock of singular terms, permitting, e.g., a description ⌜(ια)φ⌝ to be treated as a name only if we have,

$$⌜V_{(1)} \ldots V_{(m)} (E \beta)V_{(1)} \ldots V_{(n)} (\alpha)(\phi \equiv .\alpha=\beta)⌝$$

And finally we must accept the difficulties raised by,

$$(2) \qquad V_{(1)} \ldots V_{(m)} (x)(Ey)V_{(1)} \ldots V_{(n)}(x=y)^4$$

24. Belief Contexts.

Also, belief contexts may apparently be assimilated to modal contexts and interpreted along the lines of section 11. Instead of talking about possible worlds we have to talk about worlds compatible with our beliefs. Since we may believe something that is false, it may well be that our actual world is not compatible with our beliefs.

Writing 'B' for 'it is believed that',

$$(1) \qquad \ulcorner B\varphi \supset \varphi \urcorner$$

does not, therefore, in general hold. Similarly, we saw in section 22 that,

$$\ulcorner O\varphi \supset \varphi \urcorner$$

does not hold in the deontic modalities. Belief contexts and obligation contexts hence seem to behave in similar ways.

Although (1) does not hold, it seems possible to interpret belief contexts in terms of a set of worlds with are compatible with our beliefs, supplemented with our actual world. And if so, the results from the preceding chapters should carry

---

[4] The only proposed system of quantified epistemic logic is that of von Wright, presented in Chapter VI of his *Essay in Modal Logic* (1951). This system is a combination of uniform quantification theory and modal logic, and von Wright does not admit overlapping quantifiers (*An Essay in Modal Logic*, pp.48-49). By utilizing the formal analogy between 'V' and 'F' and the universal and existential quantifier, respectively, von Wright is thereby able to assimilate quantified epistemic logic to the theory of double quantification which he has worked out in "On the idea of logical truth (II)." But this restriction prevents us from seeing how (1) and (2) above would fare in his system.

over to belief contexts too. Quantification into them makes sense, on the condition that,

$$(x)(y)(x=y . \supset B(x=y))$$

The stock of singular terms must be restricted, so that in particular, a description $\ulcorner(\imath\alpha)\varphi\urcorner$ is treated as a singular term only if,

$$\ulcorner(E\beta)B(\alpha)(\varphi\equiv.\alpha=\beta)\urcorner$$

And the principle,

$$(x)(Ey)B(x=y)$$

and related difficulties parallel to those discussed in section 20 become unavoidable if one wants to quantify into belief contexts.

25. Other Contexts.

There are many types of opaque contexts whose behavior is little known, and to which the results of this thesis are therefore not easily carried over.

One important type of such contexts are the so-called act contexts.

As mentioned in section 1, already in the fourteenth century logicians were concerned about the opacity of certain verbs which relate to mental activity, like 'promises that'. Frege added a number of verbs to this list, e.g., 'say', 'hear', 'be of the opinion', 'be convinced', 'conclude', 'perceive', 'know', and 'fancy'.[5] Other verbs have been added later. Thus Quine in "Quantifiers and propositional attitudes" (1956) considers 'hunts', 'wants', 'strives', and 'wishes', in addition to 'believes', and in *Word and Object*, section 32 is devoted to such verbs. In the German intentionalist tradition (Brentano and Husserl and their followers) there is a tendency to regard all act contexts as opaque, also contexts governed by, e.g., 'sees that'.

---

[5] "Über Sinn und Bedeutung," especially p. 66 of Black's translation in *Philosophical Writings of Gottlob Frege.*

Whether and how the results of this thesis may be transferred to contexts of this type, is a question which can apparently not be answered until the formal properties of these contexts are better known.

Appendix I

Postulate Sets for the Systems of Truth-Functional Modal Logic Mentioned in the Thesis

This appendix contains nothing new, and is intended only for convenience of reference. Most of the information given in it may be found in Appendix I of Prior's *Formal Logic*, and on page 123 of Prior's *Time and Modality*. The axioms and rules below of adapted to the notation of this thesis. Throughout the appendix φ, ψ, and χ are closed sentences. '◇' is used as a sign for possibility. The other symbols peculiar to modal logic are defined as follows:

'□' for '~◇~'

⌜φ≺ψ⌝ for ⌜□(φ ⊃ ψ)⌝

⌜φ≣ψ⌝ for ⌜φ≺ψ. ψ≺φ⌝

The Lewis Systems

First presented in Appendix II of Lewis and Langford's *Symbolic Logic* (1932). S3 goes back to Lewis's *A Survey of Symbolic Logic*.[1] In the presentation below, axioms which later have been found to be redundant are left out, but Lewis's own numbering of the axioms has been retained.

---

[1] *A Survey of Symbolic Logic*, Berkeley, 1918.

Rules:

Substitution: If χ' is like χ except for containing ψ at some places where χ contains φ, then

$$\vdash \ulcorner \varphi \equiv \psi. \supset . \chi \equiv \chi'$$

Adjunction: If $\vdash \varphi$ and $\vdash \psi$, then $\vdash \ulcorner \varphi. \psi \urcorner$

Inference: If $\vdash \varphi$ and $\vdash \ulcorner \varphi \prec \psi \urcorner$, then $\vdash \psi$

Axioms:

Axioms common to all the systems:

B.1	$\ulcorner \varphi. \psi. \prec. \psi. \varphi \urcorner$
B.2	$\ulcorner \varphi. \psi. \prec. \varphi \urcorner$
B.3	$\ulcorner \varphi \prec . \varphi. \varphi \urcorner$
B.4	$\ulcorner \varphi. \psi: \chi: \prec: \varphi: \psi. \chi \urcorner$
B.6	$\ulcorner \varphi \prec \psi. \psi \prec \chi: \prec. \varphi \prec \chi \urcorner$

(In Lewis's original formulation, a sixth axiom, B.5: $\ulcorner \varphi \prec \sim\sim \varphi \urcorner$, was common to all the systems. McKinsey, in "A reduction in the number of postulates for C. I. Lewis's system of strict implication" (1934), proved, however, that this axiom is deducible from axioms B.1, B.2, B.3, and B.6.)

Axioms for:

S1:	B.1–B.6+B.7:	$\ulcorner \varphi. \varphi \prec \psi: \prec \psi \urcorner$
S2:	B.1–B.6+B.8:	$\ulcorner \Diamond(\varphi. \psi) \prec \Diamond\varphi \urcorner$
S3:	B.1–B.6+A.8:	$\ulcorner \varphi \prec \psi. \prec. \sim \Diamond\psi \prec \sim \Diamond\varphi \urcorner$
S4:	B.1–B.6+C.10:	$\ulcorner \sim\Diamond\sim\Diamond \prec \sim \Diamond\sim\varphi \urcorner$
S5:	B.1–B.6+C.11:	$\ulcorner \Diamond\varphi \prec \sim \Diamond\sim\Diamond\varphi \urcorner$
S6:	The axioms of S2+C.13:	$\ulcorner \Diamond\Diamond\varphi \urcorner$
S7:	The axioms of S3+C.13:	$\ulcorner \Diamond\Diamond\varphi \urcorner$
S8:	The axioms of S3+	$\ulcorner \sim\Diamond\sim\Diamond\Diamond\varphi \urcorner$

Gödel's System

First presented in Gödel's "Eine Interpretation des intuitionistischen Aussagenkalküls" (1933). Gödel himself proposed a system consisting of the rule RL and the three axioms A.1, A.2, and A.4 below, noticing that it was equivalent to the Lewis system S4. The singling out of Gödel's "basic" system for special consideration is, as far as I know, due to Feys (see Feys's system, below).

Gödel's basic system.

A set of axioms and rules for classical truth-functional logic supplemented by:

Rule:

RL: If ⊢φ, then ⊢⌜□φ⌝

Axioms:

A.1   ⌜□φ⊃φ⌝
A.2   ⌜□(φ⊃ψ) ⊃.□φ⊃□ψ⌝

Gödel's original system, which is equivalent to S4, is got from the above basic system by adding the axiom

A.4   ⌜□φ⊃□□φ⌝

A system equivalent to S5 is got from the basic system by adding the axiom

A.5   ⌜◇φ⊃□◇φ⌝

When Gödel's basic system is supplemented by "Brouwer's axiom"

A.3   ⌜φ⊃□◇φ⌝

a system results which in Chapter Four was found to be of special interest with regard to the distinctness of individuals and mixtures of quantifiers and modal operators. In the diagram at the end of this appendix, this system is called *Brouwer's system*.

The System of Feys

The system of Feys, presented in sections 22-26 of Feys's article "Les logiques nouvelles des modalités" (1937), is identical with Gödel's basic system. Feys, in this article, was as far as I know, the first to utilize the axiom system proposed by Gödel, and also the first to direct attention to the "basic" system got from Gödel's by deleting Gödel's axiom A.4 (above).

Rule RL above is Feys's rule 25.2, axiom A.2 above is Feys's axiom 25.3. Instead of axiom A.1 above, Feys uses $\ulcorner \varphi \supset \Diamond \varphi \urcorner$ (his 23.11).

The von Wright System

These systems were first presented in Appendix II of von Wright's *Essay in Modal Logic* (1951).

Rules:

The rules of classical sentential logic, supplemented with,

If $\vdash \ulcorner \varphi \equiv \psi \urcorner$ then $\vdash \ulcorner \Diamond \varphi = \Diamond \psi \urcorner$ (The Rule of Extensionality)

If $\vdash \varphi$ then $\vdash \ulcorner \Box \varphi \urcorner$ (The Rule of Tautology)

Axioms:

System M: A set of axioms for classical sentential logic, supplemented with,

$\ulcorner \varphi \supset \Diamond \varphi \urcorner$ (The Axiom of Possibility)

$\ulcorner \Diamond (\varphi \lor \psi) \equiv . \Diamond \varphi \lor \Diamond \psi \urcorner$ (The Axiom of Distribution)

System M': The axioms of system M +

⌜◇◇φ⊃◇φ⌝   (The First Axiom of
                                                                Reduction)

System M'': The axioms of system M +

⌜◇~◇φ⊃~◇φ⌝   (The Second Axiom of
                                                                Reduction)

Interrelations between the Systems

The following diagram, due mainly to Prior (*Time and Modality*, p.123), shows how the above systems are interrelated. An arrow from one system to another indicates that the former is stronger than the latter.

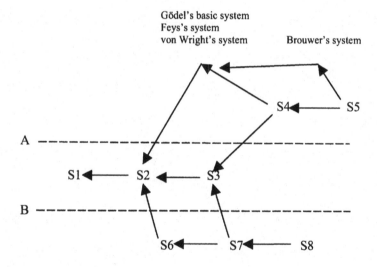

Gödel's basic system
Feys's system
von Wright's system   Brouwer's system

Systems above line A contain the rule RL (If ⊢φ then ⊢⌜□φ⌝), systems below line A do not contain it. Systems below the line B contain ⌜◇◇φ⌝ as a theorem, system above line B do not contain it. Systems above line A are incompatible with ⌜◇◇φ⌝, systems below line B are incompatible with rule RL.

The System of Church

Church's system of intensional logic, presented in "A formulation of the logic of sense and denotation" (1951), does not properly belong in this thesis, since it is not a system of modal logic in the ordinary sense (the modal operators do not attach to sentences), and since, moreover, as we shall see, the constructions of this system are apparently neither referentially nor extensionally opaque.

Nevertheless, just this latter feature of the system makes it rather interesting from our point of view, since one of the difficulties which we have tried to overcome in this thesis is that modal contexts have to be referentially transparent in order that we shall be able to quantify into them, and nevertheless must be extensionally opaque in order that modal distinction shall not collapse. The restrictions on singular terms and descriptions which we introduced in Chapter Five could apparently be avoided if modal contexts could be construed as both referentially *and* extensionally transparent.

And if Church is right, this can be done if one lets the modal operators apply to names of propositions instead of to names of truth-values.

Let us therefore consider Church's system to see how and why it works.

First, Church bases his system on Frege's theory of meaning. A basic feature of this theory is that singular terms, general terms, and closed sentences are treated on a par, as names of individual objects, classes, and truth-values, respectively, This has as a consequence that the four definitions of referential and extensional position in section 2 of this thesis (Definitions 1.a, 1.b, 2.a, and 2.b) may be fused into one:

Referential   position   of   an   expression   in   an   expression:
interchangeability of co-referential expressions *salva designatione.*

Likewise, instead of talking about referential and extensional transparency
(opacity), we may just talk about referential transparency (opacity).

If the three criteria for referential transparency given in sections 2 and 3 are
equivalent, we should therefore expect that one may quantify into a context in
Church's system only if in it *all* co-referential expression, be they singular
terms, general terms, or sentences, may be substituted for each other.

This seems indeed to be the case in Church's system. In his presentation of
the system in "A formulation of the logic of sense and denotation" (1951),
Church limits himself to setting down the axioms. But in "A formulation of the
simple theory of types" (1940), on which Church's system is based,

$$(1) \qquad x_\alpha = y_\alpha . \supset . f_{\beta\alpha} \, x_\alpha = f_{\beta\alpha} \, y_\alpha$$

(for all types α and β) are theorems (Theorems $18^{\beta\alpha}$ on page 63 of "A
formulation of the simple theory of types"). Now (1) itself cannot be a theorem
of Church's system in "A formulation of the logic of sense and denotation,"
since in this latter system formulae containing free variables may not be
asserted.[1] But the closure of (1) is apparently provable in Church's logic of
sense and denotation, for among the few theorems which Church states in "A
formulation of the logic of sense and denotation" are the following:

$$(2) \qquad (x_\alpha) \, (y_\alpha)(x_{\alpha 1})(y_{\alpha 1})(\Delta x_\alpha x_{\alpha 1} \supset: \Delta \, y_\alpha \, y_{\alpha 1} \supset . \, x_{\alpha 1} = y_{\alpha 1} \supset N(x_{\alpha 1} = y_{\alpha 1}))$$

for all types α),[2] i.e., identicals are necessarily identical, and as we saw in
section 13, the corresponding theorem in ordinary systems of modal logic

---

[1] "A formulation of the logic of sense and denotation" (1951), p. 7. In this
appendix, all references by page numbers in the text are to this article of
Church's.

[2] The dots and parentheses are mine. Church leaves most of them out on the
convention that association is to the left. Church believes that the formulae (2)
are theorems of his systems, "else they also would have been assumed as
axioms." ("A formulation of the logic of sense and denotation" (1951) p. 21).
The formulae (2) are derivable in Church's system only if one chooses his
Alternative (2), according to which one "makes the sense of A and B the same

apparently requires for its proof the universal substitutivity of identity, i.e., something rather similar to (1).

The 'α' in (1) may be any type of symbol, including e.g., 'o', the type of truth-values. So not only co-referential singular terms, but also co-referential sentences seem to be universally substitutable in Church's system.

Church's system also has another peculiarity which distinguishes it from the systems of modal logic which we have considered in this thesis: The system contains a theory of descriptions in which descriptions are treated as names and in which there are nevertheless no restrictions on the conditions that may occur in descriptions. As we saw in section 16, this would lead to a collapse of modal distinctions in ordinary systems of modal logic.

Church's system does, indeed, admit an interpretation according to which all truths are necessary, and Church observes that this may be made the basis of a relative consistency proof to the effect that his system with axiom of infinity added is consistent if the simple theory of types with axiom if infinity is consistent. (pp.21-22). The possibility of such a trivial interpretation might be avoided if one added as an axiom that there exists a proposition which is true, but not necessary. But Church refrains from adding such an axiom "not because of any doubt that it is true, but because of a doubt whether it is necessary and therefore appropriate to be assumed as an axiom of logic." (p.22).

One might think that the adding of such an axiom would perhaps render the system inconsistent, but this is apparently not the case. To see this we will now try to get a more intuitive picture of what is going on in Church's system.

Church divides his entities into types and then builds up his system with the help of one-argument functions which map one type into another, or into itself. May-argument functions are ingeniously constricted from one-argument ones.

The necessity operator, in particular, may be regarded as a function from propositions to truth-values.[3] It maps a proposition onto the truth-value truth if

---

whenever A=B is logically valid." ("A formulation of the logic of sense and denotation" (1951) p.5). Alternative (2) is the only alternative Church works out in detail, and the only of his three alternative which will be considered in this appendix. Church uses 'N' rather than '□' for necessity.

[3] Church's '$N_{o_m o_n}$' is, as indicated by the subscripts, more generally a function from truth-values, or concepts of concepts . . . of concepts of truth-values, to truth-values, or concepts of concepts . . . of concepts of truth-values. But only the 'N_{oo_1}' is interpreted as meaning logical validity or necessity (p. 19 of

the proposition is identical with the sense of an arbitrarily chosen analytic sentence, and onto the truth-value falsehood otherwise.

And it is not hard to see why the constructions of Church's system are transparent. The necessity operator is a function on a line with all other functions, and there is no reason to restrict substitution in necessity contexts any more than in other contexts. The necessity function takes only propositions as arguments, and we are free to refer to such a proposition by whichever name we wish, all that matters is that the proposition referred to remains the same.

Since the necessity operator takes only propositions as arguments, we might think that then, at least, we can only use variables taking intensions as values when we quantify into a necessity context. But this is not the case. Examples of necessity contexts which contain, e.g., free individual variables can be constructed at will. Thus, e.g., Church has suggested the following example:[4]

$$(3) \qquad N_{001} \, (f_{011} \, x_1)$$

This expression, which contains the free individual variable 'x_1', is well-formed in Church's system. If we bind this variable with a universal quantifier we get the expression,

$$(4) \qquad (x_1)(\, N_{001} \, (f_{011} \, x_1)$$

which is well-formed. The expression (4) says simply that whichever individual x_1 we choose, the proposition onto which x_1 is mapped by the function f_{011} is identical with the sense of our arbitrarily chosen analytic sentence.

Church's system hence seems to work and to be free of the many restrictions which hamper ordinary systems of quantified modal logic.

Church's article). The other 'N''s become, however, needed to express iterated modalities; cf. e.g., the formulae on p. 20 of Church's article.

[4] This example is quoted in Kemeny's review of Quine's essay "Reference and modality." Kemeny tells that the example, one of many different types allowed in the system, was suggested to him by Church.

Addendum

In the summer of 1963 I rewrote the dissertation in connection with an application for a job at the University of Oslo. In January 1966 this improved version was published in mimeographed form by Oslo University Press. The following addendum includes the main changes and additions I made in 1963.

The most important item is an addition to section 20, where I discuss necessary existence, arguing that the requirement in the original version that names and other genuine singular terms have to refer to the same object in all possible worlds was too strong. All that is needed is that they refer to the same object in all worlds where the object exists. I here used Frege's idea of letting formulas be without truth value if they contain names without a reference. The move was inspired by Arthur Prior's modal system Q in *Time and Modality* (Prior 1957). This addition affects the end of section 20 and also the many other places where I say that names and other genuine singular terms keep their reference in all possible worlds.

Apart from this the addendum contains only some supplementary observations from 1963 on the topics dealt with in the dissertation. I have sorted the additions into twelve groups:

1 **Necessary existence**
2 **More on referential vs. extensional opacity**
3 **Interpreting iterated modalities**
4 **More on the ancient and medieval discussion**
5 **Quine's slingshot**

In order to leave the original text undisturbed no marks have been put into the text. Instead the location of each change is specified below by section and page.

1. Necessary existence: names without a reference, distinguishing '□' and '~◊~'

Section 20 Other Difficulties in Quantified Modal Logic

This is the most important addition to the 1963 version.

I discuss these points more fully in: *A model theoretic approach to causal logic*. (Det Kgl. Norske Videnskabers Selskabs Skrifter. 1966. No. 2) Trondheim: Bruns Bokhandel, 1966, and in "Knowledge, identity and existence." *Theoria* 33 (1967), pp. 1-27.

Page 96. In 1963, I replaced the last paragraph with the following text:

However, quantified modal logic can be interpreted so as to avoid (1) [that is: $\Box_{(1)} \ldots \Box_{(m)} (x)(Ey) \Box_{(1)} \ldots \Box_{(n)} (x = y)$], and in several different ways. In what follows, we shall limit ourselves to considering one such solution.[1] Its basic idea is to *give up the requirement that every well-formed formula must have a truth value in every possible world.*[2] We need then no longer require, as

---

[1] The virtues of this solution become especially apparent when it is applied to the touchstone of modal ideas, the epistemic modalities. However, these modalities will not be discussed in detail in this essay; only a few basic points which relate to our theme, referential opacity, will be mentioned in section 23.

[2] This idea of letting some formulas be without a truth value in some possible worlds is reminiscent of Frege's and several others' treatment of sentences

we did in section 11, that every free variable in every formula have a reference in every possible world. We permit interpretations that in some possible worlds fail to assign any object to some free variables in our formulas; and we regard these formulas as without truth values in these worlds under these interpretations.

Given this modified interpretation,[3] we can now define validity as follows: A formula is valid if and only if there is no interpretation that makes it false.

The objects over which we quantify may now disappear as we pass form one world to one which is possible with respect to it, i.e. they do not have to exist necessarily. One consequence of this is that $\ulcorner \Box \varphi \urcorner$ is no longer equivalent to $\ulcorner \sim \Diamond \sim \varphi \urcorner$. For if φ is without a truth value in some world which is possible with respect to the actual one, and false in no such world, $\ulcorner \sim \Diamond \sim \varphi \urcorner$ will be true and yet $\ulcorner \Box \varphi \urcorner$ false. So,

3) $\ulcorner \sim \Diamond \sim \varphi \supset \Box \varphi \urcorner$

is not valid. If φ is true in every such world, however, φ will be false in none. So

4) $\ulcorner \Box \varphi \supset \sim \Diamond \sim \varphi \urcorner$

remains valid. Likewise, even if φ is not false in every such world, φ may fail to be true in any of them. So

$\ulcorner \sim \Box \sim \varphi \supset \Diamond \varphi \urcorner$

is *not* valid. If, however, φ is true in some such world, then φ is not false in all of them. So

5) $\ulcorner \sim \Box \sim \varphi \supset \Diamond \varphi \urcorner$

remains valid.

---

which contain names without a reference (cf. Frege, "Ueber Sinn und Bedeutung" (Frege 1892)). As far as I know, the first to use it in modal logic was A. N. Prior in his system Q in *Time and Modality* (Prior 1957).

[3] A full, formal definition of this notion of interpretation is given in my paper "A model theoretic approach to causal logic." Forthcoming in *Det Kongelige Norske Videnskabers Selskabs Skrifter*.

The Gödel axiom A.1 is not affected by our permitting formulas without truth values. It remains valid in the forms

A.1$_\square$ $\ulcorner \square\ \varphi \supset \varphi \urcorner$

and

A.1$_\lozenge$ $\ulcorner \varphi \supset \lozenge\ \varphi \urcorner$

as well as in the forms

A.1$_{\sim\square\sim}$ $\ulcorner \varphi \supset \sim\square\sim \varphi \urcorner$

and

A.1$_{\sim\lozenge\sim}$ $\ulcorner \sim\lozenge\sim \varphi \supset \varphi \urcorner$

That is, every formula in the following string of formulas implies every formula on its right and it implied by every formula on its left:

$$\ulcorner \square\ \varphi \urcorner \rightarrow \ulcorner \sim\lozenge\sim \varphi \urcorner \rightarrow \varphi \rightarrow \ulcorner \lozenge\ \varphi \urcorner \rightarrow \ulcorner \sim\square\sim \varphi \urcorner$$

The Gödel axiom A.2 holds in the form:

A.2$_\square$ $\ulcorner \square\ (\varphi \supset \Psi) \supset . \square\ \varphi \supset \square\Psi \urcorner$

For if here the main antecedent is true, φ and Ψ must have truth values in *all* worlds which are possible with respect to the actual one, so the situation remains the same as it was before we reinterpreted the modalities by permitting formulas without truth values. The formula

6) $\ulcorner \sim\lozenge\sim (\varphi \supset \Psi) \supset . \sim\lozenge\sim \varphi \supset \sim\lozenge\sim \Psi \urcorner$

got from A.2$_\square$ by putting '$\sim\lozenge\sim$' for '$\square$', is, however, not valid. For if Ψ is false in some world in which φ lacks truth value,

$\ulcorner \sim\lozenge\sim (\varphi \supset \Psi) \urcorner$

and

⌜~◇~ φ⌝

may both be true and yet

⌜~◇~ Ψ⌝

false. In order to prevent this, we have to require that *all free variables in* φ *occur in* Ψ.[4] For if they do, then whenever a free variable in φ lacks a reference, so does a variable in Ψ. Thereby we are insured that whenever φ is without a truth value so is Ψ. So the parallel to A.2$_\square$ which we were seeking, takes the following form:

A.2$_\diamond$ ⌜~◇~ (φ ⊃ Ψ) ⊃. ~◇~ φ ⊃ ~◇~ Ψ⌝
      provided that every free variable in φ *occurs in* Ψ.

The Gödel rule RL fares similarly, except that it remains sound in the form

RL$_\diamond$ ⊢ φ → ⊢ ⌜~◇~ φ⌝

but breaks down in the form

7)   ⊢ φ → ⊢ ⌜□ φ⌝

For although e.g. ⌜Ψ v ~ Ψ⌝ is derivable, it may fail to be true in every possible world, because in some of them it may lack truth value. This will be prevented if we make sure that φ has a truth value in every possible world, i.e. if we give the rule the form

RL$_\square$ ⊢ φ → ⊢ ⌜□ φ⌝ *provided that* φ *has no free variables.*

---

[4] Hintikka, in his review of Prior's *Time and Modality* (*Philosophical Review* 67 (1958), pp. 401-404) proposes the opposite condition, viz. That all free variables in Ψ must occur in φ (p. 402). Since Hintikka gives no proof or justification for his condition, it is hard to tell whether this is just a slip or whether Hintikka has a different application if Prior's idea in mind.

When we permit formulas without truth values, we also have to modify our theses and their corollaries in sections 12 and 14. In section 12 we found that in order to be able to say about an object in our universe of discourse that *it* has such and such properties in such and such a possible world, we had to require that there was one and only one such object in that possible world. Since we then wanted our formulas to have a truth value in every possible world, we had to require that

$$8) \qquad (x)(y)\,(x = y\,.\supset \Box\,(x = y))$$

be true. Now we no longer require our formulas to have a truth value in every possible world. However, when we say about an object in our universe of discourse that *it* has such and such properties in a certain possible world, then *if this statement has a truth value in that world*, there must be one and only one such object in that world. So, instead of (8) the weaker condition

$$9) \qquad (x)(y)\,(x = y\,.\supset {\sim}\Diamond{\sim}\,(x = y))$$

must hold. In the case of iterated modalities, considerations parallel to those which in section 12 led us to

$$10) \qquad \Box_{(1)}\ldots\Box_{(m)}\,(x)(y)\,(x = y\,.\supset\Box_{(1)}\ldots\Box_{(n)}\,(x = y))$$

$$11) \qquad \Box_{(1)}\ldots\Box_{(m)}\,(x)\,\Box_{(1)}\ldots\Box_{(n)}\,(x = x))$$

and

$$12) \qquad \Box_{(1)}\ldots\Box_{(m)}\,(x)\,(Ey)\,\Box_{(1)}\ldots\Box_{(n)}\,(x = y))$$

now make us require that, even when the notion of interpretation is modified as we have modified it in this section by permitting formulas to be without truth value in some possible worlds,

$$13) \qquad {\sim}\Diamond_{(1)}{\sim}\ldots{\sim}\Diamond_{(m)}{\sim}\,(x)(y)\,(x = y\,.\supset{\sim}\Diamond_{(1)}{\sim}\ldots{\sim}\Diamond_{(n)}{\sim}(x = y))$$

$$14) \qquad {\sim}\Diamond_{(1)}{\sim}\ldots{\sim}\Diamond_{(m)}{\sim}\,(x)\,{\sim}\Diamond_{(1)}{\sim}\ldots{\sim}\Diamond_{(n)}{\sim}(x = x))$$

$$15) \qquad {\sim}\Diamond_{(1)}{\sim}\ldots{\sim}\Diamond_{(m)}{\sim}\,(x)\,(Ey)\,{\sim}\Diamond_{(1)}{\sim}\ldots{\sim}\Diamond_{(n)}{\sim}(x = y))$$

must be true in every interpreted system of modal logic (with quantifiers and identity).

Thesis 1 of section 10 and its three corollaries must be weakened correspondingly, by putting '$\sim\Diamond\sim$' for '$\Box$' everywhere in them.

The considerations concerning distinctness of individuals in section 14 carry through with the same modifications as we just had to make in the case if identity. So even when the notion of interpretation is modified by permitting formulas to be without a truth value in some possible worlds,

16) $\sim\Diamond_{(1)}\sim\ldots\sim\Diamond_{(m)}\sim (x)(y)\,(x\neq y\,.\supset\sim\Diamond_{(1)}\sim\ldots\sim\Diamond_{(n)}\sim(x\neq y))$

must be true in every interpreted system of modal logic (with quantifiers and identity).

So also Thesis 4 and its corollary carry over in the modified form, i.e. with '$\sim\Diamond\sim$' for '$\Box$'.

Theses 2 and 3 and their corollaries, however, which concern mixtures of quantifiers and modal operators, break down when the notion of interpretation is modified the way we have done it in this section. For these theses and corollaries turn on the assumption that no object can disappear when we pass from one possible world to one that is possible with respect to it. As we have seen, this assumption need not be made if we permit formulas to be without a truth value in some possible worlds.

Earlier in this section we noted that

(4) $\ulcorner\Box\,\varphi\supset\sim\Diamond\sim\varphi\urcorner$

(6) $\ulcorner\sim\Diamond\sim(\varphi\supset\Psi)\supset.\sim\Diamond\sim\varphi\supset\sim\Diamond\sim\Psi\urcorner$

and

(7) $\vdash\varphi\to\vdash\ulcorner\Box\varphi\urcorner$

no longer remain valid if the notion of interpretation is modified so as to allow formulas to be without a truth value in some possible worlds. So for systems of modal logic which have (4) or (6) as theorems or (7) as a (primitive or derived) rule, all the theses and corollaries in sections 12 and 14 hold without modification. Since all the systems of modal logic which have been proved semantically incomplete in this essay have (4) or (6) as a theorem or (7) as a rule, these incompleteness results and proofs continue to hold even if the notion of interpretation is modified so as to permit formulas without truth values.

All three approaches to the problem of singular terms that were worked out in section 17, carry over unchanged. One might think that in approaches 2) and 3) the condition

17) $\quad \ulcorner (E\beta)\square(\alpha)(\varphi \equiv. \alpha =\beta)\urcorner$

for treating a description $\ulcorner (\iota\alpha)\varphi\urcorner$ as a name could be changed to

18) $\quad \ulcorner (E\beta){\sim}\Diamond{\sim}(\alpha)(\varphi \equiv. \alpha =\beta)\urcorner$

and similarly for iterated modalities. This is, however, not the case, as shown by the following example of two possible worlds with two objects, a and b, in the actual world and just one object, b, in a world which is possible with respect to the actual one:

Possible, non-actual world		Fb·Gb
Actual world	Fa· ~Ga	~Fb·Gb

Here '$(\iota x) Fx$' satisfies condition (18), and yet the following argument by substitutivity of identity has true premises and a false conclusion:

$\Diamond G (\iota x) Fx$
$(\iota x) Fx = a$
_____
$\Diamond Ga$

So (18) is too weak; even when we use our modified approach to the interpretation we have to keep condition (17). The reason is, of course, that whether a description has a descriptum in a possible world, H, is independent of whether the genuine singular terms with which the description is co-referential in our actual world have a reference in H. A description need not contain any free variable or genuine singular term, while the open sentence $\ulcorner (\alpha)(\varphi \equiv. \alpha =\beta)\urcorner$ in our conditions (17) and (18) on descriptions contains the free variable β. This open sentence, which should insure that $\ulcorner (\iota\alpha)\varphi\urcorner$ refers to the reference of β in every possible world, will be without a truth value in every world in which β lacks a reference. In such worlds, therefore, $\ulcorner (\iota\alpha)\varphi \urcorner$ may refer to any object whatever, without (18) ceasing to be true. To insure that the clause $\ulcorner (\alpha)(\varphi \equiv. \alpha =\beta) \urcorner$ does its job in every possible world, it therefore must be preceded by an '□', not an '~◇~'; that is, we need condition (17).

It remains only to be noted that this modification of the notion of interpretation, by making (12) invalid, saves us from the metaphysical difficulties of necessary existence that were discussed at the beginning of this section. Furthermore, it does so without requiring us to treat existence as a predicate: We no longer have to affirm or deny existence of the members of our universe of discourse, everything exists, and there is no need to use 'exists' in a sense different from that of the existential quantifier.

Section 12 The Identity of Individuals in Quantified Modal Logic

The discussion in this section and later sections must be modified in view of the discussion in the above 1963 version of section 20.

Page 38, Line 10 from below. After the period the following footnote was added in 1963:

That is, we want the question: "Does *it* have the property F in all possible worlds?" to be meaningful, answerable with yes or no, in every possible world. In section 20 we will discuss a modified notion of interpretation which does not require this.

Page 40, note 3. In the middle of the footnote, the five lines from 'Very far-off' to 'philosophers' have been expanded into:

Many of the results in this essay, particularly the semantical incompleteness results, depend of course on our notion of *interpretation* which was outlined in the preceding section. In section 20 we shall see that in order to avoid some metaphysical difficulties, it is desirable to change this notion of interpretation a little. This will weaken the theses and corollaries in sections 14 and 16, but not enough to affect any of the incompleteness results. In order to evade these, one would have to change quite radically the interpretations of the ordinary logical symbols or of the modal operators.

Section 21 Causal Modalities

Pages 98-99. The last twelve lines of page 98 and the top four lines of page 99 were replaced by:

By reasoning exactly parallel to that of section 12, it is seen that in order to quantify into contexts of natural necessity, one must require that

(1') $(x)(y)(x=y. \supset C(x=y))$

unless one permits one's formulas to be without a truth value in some possible worlds. If iterated modalities are permitted, additional 'C's will be needed in (1). From (1) we may derive

(2) $(x)(Ey)C(x=y)$

That is, every object in our actual world has to exist in every world that is compatible with the laws of nature.

In order to prevent this, one may, as we did in section 20, permit some formulas to be without a truth value in some possible worlds. Instead of (1) one then merely must require that

(1') $(x)(y)(x=y. \supset \sim A \sim (x=y))$

be valid, where '$\sim A \sim$' stands for 'it is physically possible that' ('A' for 'attainable' or 'admissible'). From (1') no undesirable consequences seem to follow, as long as one restricts one's stock of singular terms as indicated in section 20, treating, for example, a description as a singular term only if

$\ulcorner (E\ \beta) \sim A \sim (\alpha)(\varphi \equiv. \alpha = \beta) \urcorner$

Page 99, note 2. The first sentence of the footnote was replaced by:

The only proposed systems of quantified causal logic are those of Burks and Montague, proposed in "The logic of causal propositions" (1951) and "Logical necessity, physical necessity, ethics, and quantifiers" (*Inquiry* 3 (1960), pp. 259-269), respectively.

In Montague's system (1) and (2) are easily proved (by help of (A) on his page 266).

Section 22 Deontic Modalities

Page 100, lines 3-5. After the words 'section 11' the rest of the sentence was replaced by:

we might tentatively, following Hintikka in "Quantifiers in deontic logic,"[5] take the permissible to be that which is true in some world in which all one's obligations are fulfilled.

Page 101. Line 5 from below was replaced by:

be valid, unless we permit formulas to be without a truth value in some possible worlds, as we did in Section 20.

Page 103. The text on this page was replaced by:

Since as a consequence of (4) we have,

$$(5) \qquad O_{(1)} \cdots O_{(m)} \cdot (x)(Ey)\, O_{(1)} \cdots O_{(n)}\, (x = y)$$

there is good reason to modify our notion of interpretation as we did in section 20. Instead of (4) we then get

$$(4') \qquad {\sim}\, P_{(1)}{\sim} \cdots {\sim} P_{(m)}{\sim}\, (x)(y)(x{=}y. \supset\ {\sim}\, P_{(1)}{\sim} \cdots {\sim} P_{(n)}{\sim}\, (x = y))$$

from which no awkward consequences seem to follow.

Reasoning parallel to that in Chapter 5 shows further that in quantified deontic logic we have to restrict our stock of singular terms. Thus, for example, a description $\ulcorner(\iota\alpha)\varphi\urcorner$ may be treated as a name only on the condition

$$\ulcorner {\sim}\, P_{(1)}{\sim} \cdots {\sim} P_{(m)}{\sim}\, (E\,\beta)\, {\sim}\, P_{(1)}{\sim} \cdots {\sim} P_{(n)}{\sim}\, (\alpha)(\, \varphi \equiv.\, \alpha = \beta)\urcorner$$

---

[5] See the bibliography at the end of the dissertation.

(cf. section 20). Similar restrictions hold for class abstracts and other name-like expressions, as indicated in sections 17 and 20.

Section 23 Epistemic Modalities

Pages 104-105. The last seven lines of page 104 and the first two lines of page 105 were replaced by the following text, including the long footnote, where I discuss Hintikka's system of epistemic logic in his *Knowledge and Belief* (1962),[6] which was published the year after the dissertation was finished. A fuller discussion of Hintikka's interpretation of the quantifiers may be found in my article "Interpretation of quantifiers."[7]

Using the approach of section 20, we have to require that

$$(1) \qquad (x)(y)(x{=}y. \supset\ {\sim} W {\sim} (x{=}y))$$

where 'W' stands for 'its is compatible with all that one knows that'. If we permit iterated modalities,

$$(2) \qquad {\sim} W_{(1)} {\sim} \cdots {\sim} W_{(m)} {\sim} (x)(y)(x{=}y. \supset\ {\sim} W_{(1)} {\sim} \cdots {\sim} W_{(n)} {\sim} (x{=}y))$$

is required. Further we must restrict our stock of singular terms, permitting, for example, a description $\ulcorner (\iota\alpha)\varphi \urcorner$ to be treated as a name only if

$$(3) \qquad \ulcorner {\sim} W_{(1)} {\sim} \cdots {\sim} W_{(m)} {\sim} (E\,\beta) {\sim} W_{(1)} {\sim} \cdots {\sim} W_{(n)} {\sim} (\alpha)(\ \varphi \equiv. \ \alpha = \beta) \urcorner$$

(1) and (2) may seem undesirable in epistemic logic. However, in view of restriction (3) and section 20, they are not.[8]

---

[6] Ithaca, NY: Cornell, 1962.

[7] In B. van Rootselaar and J. F. Staal, eds., *Logic, Methodology and Philosophy of Science.*(Proceedings of the Third International Congress for Logic, Methodology and the Philosophy of Science, Amsterdam 1967) Amsterdam: North-Holland, 1968, pp. 271-81.

[8] The only proposed systems of quantified epistemic logic are those of von Wright and Hintikka.

Section 24 Belief Contexts

Page 106. The last part of the section, starting after the words 'belief contexts too' on line 1, was replaced by the following text:

If we use the approach of section 20, quantification into belief contexts makes sense if

(2) $(x)(y)(x=y. \supset \sim D \sim (x=y))$

where 'D' stands for 'its is compatible with what is believed that'. The stock of singular terms must be restricted, so that a description $\ulcorner(\iota\alpha)\varphi\urcorner$ is treated as a name only if

---

Von Wright's systems, EV and VE+EV, presented in Chapter VI of *An Essay in Modal Logic* (Von Wright 1951), are combinations of uniform quantification theory and epistemic logic, and von Wright does not permit overlapping quantifiers (*An Essay in Modal Logic*, pp. 48-49). By utilizing the formal analogy between his 'V' and '~F' (it is not falsified that', corresponding to our 'W' above), and the universal and existential quantifier, respectively, von Wright is thereby able to assimilate quantified epistemic logic to the theory of double quantification which he has worked out in "On the idea of logical truth (II) (1950)." But due to this restriction, requirements (1) and (2) above cannot be formulated in his systems.

Hintikka's system, which was presented in the last chapter of *Knowledge and Belief* (1962), combines epistemic operators and general quantification theory. However, in order to get around the difficulties connected with quantifying into epistemic contexts, which we believe to have solved above, Hintikka modifies his quantification theory and its interpretation in such a way that what results is not what in the semantical considerations of this essay we understand by quantification theory. The difference comes most strikingly to the surface in the reading of the quantifier. Thus, for example, (using with Hintikka 'K$_a$' for 'a knows that', and restricting the universe to men): '$(x)K_a(\ldots x \ldots)$' does not read, as we should expect, 'of each man a knows that . . .' but 'of each man known to a, a knows that . . .' (cf., for example, Hintikka's page 164).

(3) $\ulcorner (E\ \beta) \sim D \sim (\alpha)(\ \varphi \equiv .\ \alpha = \beta)\urcorner$

The generalizations of (2) and (3) for iterated modalities are obvious.

2. More on referential vs. extensional opacity

Section 2 Criteria for Referential and Extensional Opacity

Page 5, line 12 from below. To make the exposition clearer, the following definitions were added:

> Constructions which are not {referentially/extensionally} transparent will be called {referentially/extensionally} opaque. Instead of speaking of a {refer-entially/extensionally} transparent construction we will often for short speak of a {referential/ extensional} construction. A {referentially/ extensionally} opaque construction will similarly often for short be called a {non-referential/non-extensional} construction.

Section 5 Interrelations between Referential Opacity and Extensional Opacity

Pages 8-10. The whole section was rewritten and improved:

Interrelations between Referential Opacity and Extensional Opacity[9]

One may prove that

> 1.   *Every extensional construction on general terms or sentences is referential*

and

---

[9] The results in this section were proved independently by Quine and presented in his course *Philosophy of Language* on April 10, 1961, only a few days after this thesis was submitted, and before it had been read by anyone.

2. *Every referential construction on singular terms is extensional.*

Proof of 1:

Let

1) φ be a construction on general terms or sentences
2) φ be extensional
3) μ be a singular term and $\ulcorner \ldots \mu \ldots \urcorner$ and ingredient of the construction φ

Let further

4) $\ulcorner \mu = \upsilon \urcorner$ be true

and

5) the position of μ be referential.

Then, by 1), 4), and 5)

6) the expressions $\ulcorner \ldots \mu \ldots \urcorner$ and $\ulcorner \ldots \upsilon \ldots \urcorner$ co-extensional.

Since, by (i) of section 2 and 2) above, $\ulcorner \ldots \mu \ldots \urcorner$ is in extensional position in φ, we may by 6) conclude that

7) $\ulcorner \varphi(\ldots \mu \ldots) \urcorner$ and $\ulcorner \varphi(\ldots \upsilon \ldots) \urcorner$ are co-extensive or co-referential.

Hence the position of μ in $\ulcorner \varphi(\ldots \mu \ldots) \urcorner$ is referential, and the construction is referential.

The proof of 2 is parallel, with 'singular term' for 'general term or sentence' and conversely, 'referential' for 'extensional' and $\ulcorner \mu \equiv \upsilon \urcorner$ for $\ulcorner \mu = \upsilon \urcorner$.

In virtue of 1, all truth-functional and quantificational constructions, being extensional constructions on general terms or sentences, are referentially transparent.

If one treats all singular terms as descriptions, e.g. by insisting on the "primacy of predicates", one may also prove that

3. Every extensional construction on singular terms is referential.

Proof:

Let

1) φ be a construction on singular terms
2) φ be extensional
3) $\ulcorner (\iota\alpha) \Psi \urcorner$ be a singular terms which is an ingredient of the construction φ
4) $\ulcorner (\iota\alpha) \Psi = (\iota\alpha) \Psi' \urcorner$ be true

By 4), Ψ and Ψ' are co-extensional, both being true of one and the same object. They are both in extensional position in $\ulcorner (\iota\alpha)\ \Psi\urcorner$ and $\ulcorner (\iota\alpha)\ \Psi'\ \urcorner$, respectively, and hence, by 2), in $\ulcorner \phi(\iota\alpha)\ \Psi\urcorner$ and $\ulcorner \phi\ (\iota\alpha)\ \Psi'\ \urcorner$, respectively. I.e. $\ulcorner \phi\ (\iota\alpha)\ \Psi\urcorner$ and $\ulcorner \phi\ (\iota\alpha)\ \Psi'\urcorner$ are co-extensional and co-referential.

Q.E.D.

The only way, therefore, of obtaining a construction which is extensional but not referential is to hold that there are some singular terms which cannot be treated as descriptions and which furthermore can flank identity signs in true identity statements without obeying the law of substitutivity of identity.

An example of such a construction, due to Quine,[10] is the following: Consider a language in which all general terms or sentences are proceeded by a '□' whenever they occur in a singular term. Let us further suppose that this language, e.g., contains some singular term μ which contains no general term or sentence, and some other singular term υ, such that $\ulcorner \mu = \upsilon\urcorner$ is true and yet $\ulcorner \square\ (\mu = \upsilon)\urcorner$ is false. In such a language the Carnapian construction $\ulcorner \alpha \equiv \beta\urcorner$ ($\ulcorner \square(\alpha = \beta)\urcorner$) is extensional, but not referential.

Constructions like this have, I believe, never been used or seriously proposed. They don't mix with quantification: in Chapters III-V it will be argued that non-referential constructions cannot be quantified into. And worse, singular terms of the type required seem rather questionable philosophically (cf. section 17).

Only one case remains: referential constructions on general terms or sentences. If we could prove that all such constructions are extensional, there would be little need to distinguish referential and extensional opacity, positions and constructions.

3. Interpreting iterated modalities

Section 11 Interpreting Quantified Modal Logic

Page 30. Footnote 26 was replaced by the longer footnote:

---

[10] This example was proposed by Quine in his course Philosophy of Language on April 10, 1961, and has been inserted in this essay later.

This idea, which probably is the most important innovation in the interpretation of modal logic since Leibniz, is due to Jaakko Hintikka and Stig Kanger. It is clearly stated in Hintikka's 1957 papers "Quantifiers in deontic logic," e.g. on page 13, and "Modality as referential multiplicity" (pp. 61-62),[11] and in Kanger's book *Provability in Logic*,[12] pp. 33 ff. The idea was anticipated in work by Jónsson and Tarski on Boolean algebra in 1951.[13] Later, Saul Kripke got the idea independently and extended it (1959) so as to make it applicable to the Lewis systems S2 and S3. (Cf. his "Semantical analysis of modal logic (abstract) (1959)[14] and "Semantical analysis of modal logic I: Normal modal propositional calculi" (1963),[15] p. 69, n. 2.). I am indebted to Hintikka's 1957 papers and to conversations with Kripke for the interpretation of iterated modalities outlined in this section.

4. More on the ancient and medieval discussion of these topics

Section 1 Introduction

Page 1. In the 1963 version more historical information is included: In line 7, after the period, the following is added:

Aristotle, in *De Sophisticis Elenchis* 23 (179b1-3) discussed the same difficulty about the veiled Coriscus. The third main school of ancient logicians, the Stoics, with their highly developed Frege-type semantics, might have been expected to be especially sensitive to the semantical and logical problems of referential opacity, but the little evidence we have,

---

[11] See the bibliography at the end of this dissertation.

[12] Stockholm Studies in Philosophy, Vol. 1, Stockholm: Almqvist and Wiksell, 1957.

[13] B. Jónsson and A. Tarski, "Boolean algebra with operators," *American Journal of Mathematics*, 73 (1951), pp. 891-939 and 74 (1952), pp. 127-162.

[14] *Journal of Symbolic Logic* 24 (1959), pp. 323-324.

[15] *Zeitschrift für mathematische Logik und Grundlagen der Mathematik* 9 (1963), pp. 67-96.

indicates that they were not, possibly because the principle of substitutivity of identity was apparently not known to them.[16]

During the Middle Ages, Aristotle's modal logic and some of his sophisms were a starting point for further discussion of the problem. Beginning approximately with Peter of Spain, Aristotle's distinction between composite and divisive sense started to gain prominence.

Page 1, Footnote 2 was replaced by the following, fuller, footnote:

The difference between the notional and the relational sense is the difference between rendering

Ernest is hunting lions

as

(Ex)(x is a lion · Ernest strives that Ernest finds x) [the relational sense]

and as

Ernest strives that (Ex)(x is a lion · Ernest finds x) [the notional sense]

For this distinction, see Quine, "Quantifiers and prepositional attitudes" (Quine 1956).

That there are similarities between the logical modalities and certain verbs had been noted by several others before, among them St. Anselm and the remarkable "Pseudo-Scotus". Cf. e.g. W. and M. Kneale, *The Development of Logic*. Oxford: Clarendon Press, 1962, p. 243.

Heytesbury and Billingham are discussed in Curtis Wilson, *William Heytesbury: Medieval Logic and the Rise of Mathematical Physics*. Madison: University of Wisconsin Press, 1956. A manuscript of Billingham's *De Sensu Composito et Diviso* (Paris, Bibliothèque Nationale, Fonds Latin, MS 14715, foll. 79ra-82rb) has been edited and translated by

---

[16] Cf. Benson Mates, *Stoic Logic*, Berkeley and Los Angeles: University of California Press, 1953, 2. printing 1961, p. 21. However, even Aristotle stated most of the basic properties of identity, among them the indiscernability of identicals, which is relevant here (*Topica* vii, I, 152^{a}30)

Waud H. Kracke, as part of his B.A. thesis, written under the guidance of Professor John Murdoch: *Composite and Divisive Sense: The Structure and Content of a Medieval System of Logic* (Harvard, 1961).

5. Quine's slingshot

Section 10 Difficulties Relating to Quantification into Modal Contexts

Pages 28-29. The presentation of Quine's argument was slightly expanded, to make it easier to follow. The last three lines of page 28 and the rest of the section were replaced by:

Let 'p' stand for any true sentence, let x be any object in our purified universe of discourse, and let w = x. Then

12) $(y) (p \cdot y = w .\equiv. y=x)$

13) $(y) (y = w .\equiv. y=x)$

Introducing 'p· ① = w' for 'F ① ' and ' ① = w' for 'G ① ' in (ii) one gets:

14) $(y) (p \cdot y = w .\equiv. y =x) \cdot (y) (y = w .\equiv. y = x) .\supset \Box (y) (p \cdot y = w .\equiv. y = w)$

which, together with 12) and 13) implies:

15) $\Box (y) (p \cdot y = w .\equiv. y = w)$

But the quantification in 15) implies in particular 'p·w = w .≡. w = w', which in turn implies 'p', so from 15) we conclude:

$\Box p$

Since in this proof nothing is assumed about the objects over which we quantify, restricting the values of one's variables to intensional objects does not save one from this collapse of modal distinctions. So, unless quantification into modal contexts can be interpreted without assuming (ii), the prospects for a quantified modal logic are very gloomy indeed.

6. Aristotelian essentialism

Section 10 Difficulties Relating to Quantification into Modal Contexts

Page 28, line 13. To make more explicit what is meant by 'Aristotelian essentialism' the following lines were added at the end of the sentence:

> . . . that is the doctrine that some of the attributes of a thing are essential to it, necessary of the thing regardless of how we refer to it, while other attributes are accidental to it. For example, a man is essentially rational, not merely *qua* man, but regardless of how we refer to him.

7. Intensional ontology

Section 19 Examination of the Difficulties Surveyed in Section 10

Page 89. To emphasize the irrelevance of an intensional ontology, the first five lines were replaced by:

> To restrict the range of one's variables to intensional objects, as suggested by Church, is not a way out. As noted at the end of section 10 (cf. also section16), the collapse of modal distinctions is independent of the kind of entities over which one quantifies. Church may, however, have had in mind his "Logic of sense and denotation," which appeared a few years later, and this system works (see Appendix II).

8. Smullyan and Fitch

Section 18 Substitutivity of Identity and Other Types of Inference
Turning on Singular Terms

Page 83. The discussion of Smullyan's approach to quantification into modal
contexts was expanded. Line 4 was replaced by two pages, which here are
summarized:

It is shown that Smullyan is led astray because he characterizes the
following argument form as valid:

$(x)\Box Fx$
$\Box(E!)(\iota x)\varphi x$
$\therefore \Box F(\iota x)\varphi x$

This argument form is not valid, as can be seen from the following
example of this form:

$(x)\Box(x=9. \supset \Box(x=9))$
$\Box(E!)(\iota x)$(there are exactly x planets)
$\therefore \Box[(\iota x)$(there are exactly x planets) = 9. $\supset$
$\Box((\iota x)$(there are exactly x planets)=9)]

The first premise is true in any interpreted system of modal logic, we saw in
section 12. It is also easily proved from the assumptions Smullyan makes in
his paper. Also the second premise is true, for although 'there are exactly x
planets' is true of one number, 9, in our actual world and of other numbers
in other possible worlds, it is presumably true of one and only one number
in each possible world (0 being a number). The conclusion, however, is
false. So Smullyan's argument form is not valid.

Page 83-84. The last twelve lines on page 83 and the top two paragraphs on
page 84 were replaced by:

Condition (4) $[\Box(E\alpha)(\varphi x \equiv_x x \in \alpha)]$is weaker; the corresponding condition
for descriptions,

(8) $\ulcorner \Box (E\beta)(\alpha)(\varphi \equiv .\alpha = \beta)\urcorner$

is as we just saw, too weak. Smullyan was probably led to (4) by his unfortunate examples of inferences turning on descriptions and his resulting fallacious argument form. Thus he passes from his discussion of descriptions to his discussion of class abstracts by saying (p. 113): "Since we are here concerned to combine modalities with a logic which assumes the existence of classes, it appears natural to stipulate" axiom (4). His error concerning descriptions has perhaps been corroborated by his observation that for class abstracts (5) and (6) go poorly together.

There is, however, no need to assume (4).

Page 84. The following footnote was appended to the fourth line from the bottom of the text:

In a letter of August 28, 1961, Professor Fitch has informed me that he spotted this error himself shortly after his paper appeared and that in all or nearly all the reprints he sent out in 1949 the condition was corrected to

'$\Box$ E! (ιx) fx and fx $\prec_x \Box$ (fx)'.

9. Carnap

Section 13 The Identity of Individuals in the Proposed Systems of Quantified Modal Logic

Page 48, line 2. In the reference to Carnap's *Meaning and Necessity* the following page reference was added: 'cf. also pp. 111 ff.'.

The next three sentences up to and including ". . . soothe these doubts" were expanded into:

With this one may agree. But there is a far more controversial link in Carnap's argument. Carnap argues that since it is not possible for a predicator in an interpreted language to possess only an extension and not an intension, it is not possible for such a predicator "to *refer* only to a class and not a property" (p. 199, italics mine). Hence, according to Carnap, the universe of discourse of a language, which comprises all that is *referred to* in the language, contains both extensions and intensions, irrespective of whether the language is extensional or intensional. Carnap is the only philosopher I know who holds that expressions in ordinary extensional contexts *refer* to their intensions. Yet Carnap holds that 'possessing an intension' and 'referring to a property' are merely variant manners of speaking, and he even seems to regard the latter manner as more customary: "... to possess an intension or, in customary terms, to refer to ... a property" (p. 199).

10. Myhill

Section 15 The Distinctness of Individuals

Page 66, line 6 from below. After 'Carnap only' the following discussion of Myhill was added as a footnote:

Myhill, however, in his "Problems arising in the formalization of intensional logic" (1958),[17] sets forth a System V which has all the axioms and rules of S5, yet, unaware that (6) is a theorem of his system, Myhill rejects (6) and even gives an elaborate argument to show that (6) conflicts with intuitions which to him seem self-evidently true (p. 80). Myhill's radically conflicting intuitions and arguments give little credence to his conclusion: "What we claim to have demonstrated in this paper is merely that, *if such a discipline* [a logic of extensions] *exists at all*, it must number among its theorems all those listed by us as axioms of System V" (p. 83). [This inconsistency of Myhill's has been noted also by E. J. Lemmon in his paper "Quantified S4 and the Barcan formula," delivered at the twenty-sixth annual meeting of the Association for Symbolic Logic on December 27, 1961 in Atlantic City, NJ.][18]

---

[17] *Logique et Analyse* 1 (1958), pp. 74-83.

[18] Abstract received November 3, 1961, published October 1962 in *Journal of Symbolic Logic* 25 (1960), pp. 391-392.

11. Systems of quantified modal logic

Section 9 Systems of Quantified Modal Logic

Page 23. After line 5, the following reference was added:

> In 1960, Richard Montague published a system of quantified modal logic that he had originally proposed in a talk in 1956 ("Logical necessity, physical necessity, ethics, and quantifiers." *Inquiry* 3 (1960), pp. 259-269). The system, which is essentially based on S5, can be interpreted as a system of alethic, physical or deontic modality. Montague then shows how the system can be given three model theoretic interpretations, one for each of the three types of modality.

12. Minor points of clarification and correction

Section 4 Examples, and Further Characteristics of Referential and Extensional Opacity

Page 7. In the last line, after the period, the following was inserted:

> while they are, of course, referential again within the narrower contexts 'Tully denounced x', and 'Tully was a Roman'.

Section 7 Unquantified Modal Logic

Page 14. The following words were added at the beginning of the section:

> In most systems of modal logic

Section 8 Quantified Modal Logic

Page 16. In the fourth paragraph in footnote 7, the opening phrase, 'These two remarks by Church have' was replaced by:

The first of Church's remarks has

Section 11 Interpreting Quantified Modal Logic

Pages 29-30. In the example the terms 'Ka' and 'Kb' were deleted, since they are not needed.

Page 33, line 5. The page number has been added to the reference to Becker: p. 513.

Section 16 Definite Descriptions in Modal Logic

Page 70. The following line was added after line (4') in the proof, to make the proof conform to the inference rules:

$$(x)(x=z\cdot p .\equiv. x=y)$$

Appendix I Postulate Sets

Page 110, bottom. The following footnote was inserted:

(In Lewis' original formulation of S3, after correction by Emil Post, S3 had the axioms B1, B3, B5, B6, and the axioms A2: ⌐Ψ. φ . ≺ . φ⌐ instead of B2, A4: ⌐φ : Ψ .χ: ≺ : Ψ: φ. χ⌐ instead of B4, and A7: ⌐~◊ φ ≺ ~ φ⌐ instead of B7. However, the resulting set of axioms is equivalent to the axiom set given above.)

Page 113, bottom. Footnote 20 on page 21, on Gödel's basic system, was moved to here, where it naturally belongs.

Appendix II The System of Church

Page 117. Finally, the following lines were appended at the end of footnote 2:

Church's alternative (0) has been proved to be syntactically inconsistent in Myhill's "Problems arising in the formalization of intensional logic."[19]

---

[19] *Logique et Analyse* 1 (1958), pp. 74-83), p. 82.

Bibliography

Anderson, Alan Ross. Review of Prior's "Modality and quantification in S5." In *Journal of Symbolic Logic* 22 (1957), p. 91.

Barcan, Ruth C. "A Functional Calculus of First Order Based on Strict Implication." *Journal of Symbolic Logic* 11 (1946), pp. 1-16.

————. "The Deduction Theorem in a Functional Calculus of First Order Based on Strict Implication." *Journal of Symbolic Logic* 11 (1946), pp. 115-118.

————. "The Identity of Individuals in a Strict Functional Calculus of Second Order." *Journal of Symbolic Logic* 12 (1947), pp. 12-15.

————. Review of Smullyan's "Modality and Description." *Journal of Symbolic Logic* 13 (1948), 149-150.

Bayart, Arnould. "La correction de la logique modale de premier et second ordre S5." *Logique et Analyse* 1 (1958), pp. 28-44.

————. "Quasi-adéquation de la logique modale de second ordre S5 et adéquation de la logique modale de premier ordre S5." *Logique et Analyse* 2 (1959), pp. 99-121.

Becker, Oskar. "Zur Logik der Modalitäten." *Jahrbuch für Philosophie und phänomenologische Forschung* 11 (1930), 497-548.

Bernays, Paul. Review of Carnap's "Modalities and Quantification." *Journal of Symbolic Logic* 13 (1948), 218-219.

Bochenski, Innocentius. *Ancient Formal Logic.* Amsterdam: North-Holland Publishing Company, 1951.

Burks, Arthur W. "The Logic of Causal Propositions." *Mind* 60 (1951), pp. 363-382.

Carnap, Rudolf. "Modalities and Quantification." *Journal of Symbolic Logic* 11 (1946), pp. 33-64.

———. *Meaning and Necessity.* Chicago: University, 1947. 2nd ed., with supplements, 1956.

Church, Alonzo. "A Formulation of the Simple Theory of Types." *Journal of Symbolic Logic* 5 (1940), pp. 56-68.

———. Review of Quine's "Whitehead and the Rise of Modern Logic." *Journal of Symbolic Logic* 7 (1942), pp. 100-101.

———. Review of Quine's "Notes on Existence and Necessity." *Journal of Symbolic Logic* 8 (1943), pp. 45-47.

———. "A Formulation of the Logic of Sense and Denotation." Abstract. *Journal of Symbolic Logic* 11 (1946), p. 31.

———. Review of Lewis and Langford, *Symbolic Logic*, 2nd ed. *Journal of Symbolic Logic* 16 (1951), p. 225.

———. "A Formulation of the Logic of Sense and Denotation." In Henle, Kallen, and Langer, pp. 3-24.

Feigl, Herbert and Wilfred Sellars, eds. *Readings in Philosophical Analysis.* New York: Appleton-Century-Crofts, 1949.

Feys, Robert. "Les logiques nouvelles des modalités." *Revue néoscholastique de Philosophie* 40 (1937), pp. 517-553, and 41 (1938), pp. 217-252.

———. "Les systèmes formalises des modalités aristoteliciennes." *Revue philosophique de Louvain* 48 (1950), pp. 478-509.

Fitch, Frederic B. "Intuitionistic Modal Logic with Quantifiers." *Portugaliae Mathematica* 7 (1948), pp. 113-118.

————. "The Problem of the Morning Star and the Evening Star." *Philosophy of Science* 16 (1949), pp. 137-141.

————. *Symbolic Logic*. New York: Ronald Press, 1952.

Fraenkel, Abraham Adolf. "The Relation of Equality in Deductive Systems." *Proceedings of the Tenth International Congress of Philosophy* (Amsterdam, 1948), pp. 752-755.

Frege, Gottlob. "Ueber Sinn und Bedeutung." *Zeitschrift für Philosophie und philosophische Kritik* 100 (1892), pp. 25-50.

————. *Philosophical Writings*. Peter Geach and Max Black, eds. Oxford: Blackwell, 1952.

Gödel, Kurt. "Eine Interpretation des intuitionistischen Aussagenkalküls." *Ergebnisse eines mathematischen Kolloquiums* Heft 4 (1933), pp. 39-40.

Hallden, Sören. "On the Semantic Non-completeness of Certain Lewis Calculi." *Journal of Symbolic Logic* 16 (1951), pp. 127-129.

Henkin, Leon. *The Completeness of Formal Systems*. Unpublished thesis, Princeton University, 1947.

————. "The Completeness of Formal Systems." Abstract. *Journal of Symbolic Logic* 13 (1948), p. 61.

————. "The Completeness of the First Order Functional Calculus." *Journal of Symbolic Logic* 14 (1949), pp. 159-166.

————. "Completeness in the Theory of Types." *Journal of Symbolic Logic* 15 (1950), pp. 81-91.

————. Review of Rasiowa's "Algebraic Treatment of the Functional Calculi of Heyting and Lewis." *Journal of Symbolic Logic* 18 (1953), pp. 72-73.

————. Review of Rasiowa and Sikorski's "Algebraic Treatment of the Notion of Satisfiability." *Journal of Symbolic Logic* 20 (1955), pp. 78-80.

————. Review of Rasiowa and Sikorski's "On Existential Theorems in Non-classical Functional Calculi." *Journal of Symbolic Logic* 20 (1955), p. 80.

Henle, Paul, H. M. Kallen, and S. K. Langer, eds. *Structure, Method, and Meaning: Essays in Honor of H. M. Sheffer.* New York: Liberal Arts, 1951.

Hilbert, David and Paul Bernays. *Grundlagen der Mathematik*, volume 1. Berlin: Springer, 1934.

Hintikka, K. Jaakko J. "Modality as Referential Multiplicity." *Ajatus* 20 (1957), pp. 49-64.

————. "Quantifiers in Deontic Logic." *Societas Scientiarum Fennica, Commentationes Humanarum Literarum* 23, no. 4. Helsinki, 1957.

Kemeny, John G. Review of Quine's "Reference and Modality," essay VIII in *From a Logical Point of View. Journal of Symbolic Logic* 19 (1954), pp. 137-138.

Kripke, Saul A. "A Completeness Theorem in Modal Logic." *Journal of Symbolic Logic* 24 (1959), 1-14.

Lewis, Clarence Irving and Cooper Harold Langford. *Symbolic Logic*, New York: Century, 1932. 2nd edition with appendix by C. I. Lewis. New York: Dover, 1959.

Linsky, Leonard, ed. *Semantics and the Philosophy of Language.* Urbana: University of Illinois Press, 1952.

McKinsey, John Charles Chenoweth. "A Reduction in the number of Postulates for C. I. Lewis' System of Strict Implication." *Bulletin of the American Mathematical Society* 40 (1934), pp. 425-427.

McKinsey, J. C. C. and Alfred Tarski. "Some Theorems about the Sentential Calculi of Lewis and Heyting." *Journal of Symbolic Logic* 13 (1948), pp. 1-15.

Parry, William Tuthill. "Modalities in the *Survey* System of Strict Implication." *Journal of Symbolic Logic* 4 (1939), pp. 137-154.

Prior, Arthur Norman. *Formal Logic.* Oxford: Clarendon, 1955.

————. "Modality and Quantification in S5." *Journal of Symbolic Logic* 21 (1956), pp. 60-62.

———. *Time and Modality*. Oxford: Clarendon, 1957.

Quine, Willard van Orman. *Mathematical Logic*. New York: Norton, 1940. Revised edition, Cambridge, MA: Harvard University Press, 1951.

———. "Whitehead and the Rise of Modern Logic." In Schilpp, *The Philosophy of Alfred North Whitehead*, pp. 125-163.

———. Review of Russell's *An Inquiry into Meaning and Truth*. *Journal of Symbolic Logic* 6 (1941), pp. 29-30.

———. "Notes on Existence and Necessity." *Journal of Philosophy* 40 (1943), pp. 113-127.

———. Review of Barcan's "A Functional Calculus of First Order Based on Strict Implication." *Journal of Symbolic Logic* 11 (1946), pp. 96-97.

———. Review of Barcan's "The Deduction Theorem in a Functional Calculus of First Order Based on Strict Implication." *Journal of Symbolic Logic* 12 (1947), p. 95.

———. Review of Barcan's "The Identity of Individuals in a Strict Functional Calculus of Second Order." *Journal of Symbolic Logic* 12 (1947), pp. 95-96.

———. "The Problem of Interpreting Modal Logic." *Journal of Symbolic Logic* 12 (1947), pp. 43-48.

———. Review of Fraenkel's "The Relation of Equality in Deductive Systems." *Journal of Symbolic Logic* 14 (1949), p. 130.

———. *From a Logical Point of View*. Cambridge, MA: Harvard University Press, 1953.

———. "Three Grades of Modal Involvement." *Proceedings of the Eleventh International Congress of Philosophy* (Brussels, 1953), volume 14, pp. 65-81.

———. "Quantifiers and Propositional Attitudes." *Journal of Philosophy* 53 (1956), pp. 177-187.

———. *Word and Object*. Cambridge, MA, and New York: MIT Press and Wiley, 1960.

Rasiowa, Helena. "Algebraic Treatment of the Functional Calculi of Heyting and Lewis." *Fundamenta Mathematicae* 38 (1952), pp. 99-126.

Rasiowa, H., and R. Sikorski. "Algebraic Treatment of the Notion of Satisfiability." *Fundamenta Mathematicae* 40 (1953), pp. 62-95.

————. "On Existential Theorems in Non-classical Functional Calculi." *Fundamenta Mathematicae* 41 (1954), pp. 21-28.

Russell, Bertrand. *Mysticism and Logic and Other Essays*. London: Allen and Unwin, 1918.

Schilpp, Paul Arthur, ed. *The Philosophy of Alfred North Whitehead*, Evanston, IL: Northwestern University Press, 1941. 2nd edition. New York: Tudor, 1951.

Smullyan, Arthur Francis. Review of Quine's "The Problem of Interpreting Modal Logic." *Journal of Symbolic Logic* 12 (1947), pp. 139-141.

————. "Modality and Description." *Journal of Symbolic Logic* 13 (1948), pp. 31-37.

Sobocinski, Boleslaw. "Note on a Modal System of Feys-von Wright." *Journal of Computing Systems* 1 (1953), pp. 171-178.

von Wright, Georg Henrik. "On the Idea of Logical Trugh (I)." *Societas Scientiarum Fennica, Commentationes Physico-mathematicae* 14, no. 4, 1948. Reprinted in von Wright, *Logical Studies*.

————. "On the Idea of Logical Truth (II)." *Societas Scientiarum Fennica, CommentationesPhysico-Mathematicae* 15, no. 10, 1950.

————. *An Essay in Modal Logic*. Amsterdam: North-Holland Publishing Company, 1951.

————. *Logical Studies*. London: Routledge and Kegan Paul, 1957.

Whitehead, Alfred North and Bertrand Russell. *Principia Mathematica*, volume 1. Cambridge: Cambridge University Press, 1910. 2nd edition, 1925.

Wilson, Neil L. *The Concept of Language*. Toronto: University of Toronto Press, 1959.

Index

Printed in the United States
by Baker & Taylor Publisher Services